Research Notes in Mathematics

Main Editors
A. Jeffrey, University of Newcastle-upon-Tyne
R. G. Douglas, State University of New York at Stony Brook

Editorial Board
F. F. Bonsall, University of Edinburgh
H. Brezis, Université de Paris
G. Fichera, Università di Roma
R. P. Gilbert, University of Delaware
K. Kirchgässner, Universität Stuttgart
R. E. Meyer, University of Wisconsin-Madison
J. Nitsche, Universität Freiburg
L. E. Payne, Cornell University
G. F. Roach, University of Strathclyde
I. N. Stewart, University of Warwick
S. J. Taylor, University of Virginia

Submission of proposals for consideration
Suggestions for publication, in the form of outlines and representative samples, are invited by the editorial board for assessment. Intending authors should contact either the main editor or another member of the editorial board, citing the relevant AMS subject classifications. Refereeing is by members of the board and other mathematical authorities in the topic concerned, located throughout the world.

Preparation of accepted manuscripts
On acceptance of a proposal, the publisher will supply full instructions for the preparation of manuscripts in a form suitable for direct photo-lithographic reproduction. Specially printed grid sheets are provided and a contribution is offered by the publisher towards the cost of typing.

Illustrations should be prepared by the authors, ready for direct reproduction without further improvement. The use of hand-drawn symbols should be avoided wherever possible, in order to maintain maximum clarity of the text.

The publisher will be pleased to give any guidance necessary during the preparation of a typescript, and will be happy to answer any queries.

Important note
In order to avoid later retyping, intending authors are strongly urged not to begin final preparation of a typescript before receiving the publisher's guidelines and special paper. In this way it is hoped to preserve the uniform appearance of the series.

Titles in this series

56 An approach to algebraic K-theory
A J Berrick
57 Convex analysis and optimization
J-P Aubin and R B Vintner
58 Convex analysis with applications in the differentiation of convex functions
J R Giles
59 Weak and variational methods for moving boundary problems
C M Elliott and J R Ockendon
60 Nonlinear partial differential equations and their applications: Collège de France Seminar. Volume II
H Brezis and J L Lions
61 Singular systems of differential equations II
S L Campbell
62 Rates of convergence in the central limit theorem
Peter Hall
63 Solution of differential equations by means of one-parameter groups
J M Hill
64 Hankel operators on Hilbert space
S C Power
65 Schrödinger-type operators with continuous spectra
M S P Eastham and H Kalf
66 Recent applications of generalized inverses
S L Campbell
67 Riesz and Fredholm theory in Banach algebra
B A Barnes, G J Murphy, M R F Smyth and T T West
68 Evolution equations and their applications
F Kappel and W Schappacher
69 Generalized solutions of Hamilton-Jacobi equations
P L Lions
70 Nonlinear partial differential equations and their applications: Collège de France Seminar. Volume III
H Brezis and J L Lions
71 Spectral theory and wave operators for the Schrödinger equation
A M Berthier
72 Approximation of Hilbert space operators I
D A Herrero
73 Vector valued Nevanlinna Theory
H J W Ziegler
74 Instability, nonexistence and weighted energy methods in fluid dynamics and related theories
B Straughan
75 Local bifurcation and symmetry
A Vanderbauwhede
76 Clifford analysis
F Brackx, R Delanghe and F Sommen
77 Nonlinear equivalence, reduction of PDEs to ODEs and fast convergent numerical methods
E E Rosinger
78 Free boundary problems, theory and applications. Volume I
A Fasano and M Primicerio
79 Free boundary problems, theory and applications. Volume II
A Fasano and M Primicerio
80 Symplectic geometry
A Crumeyrolle and J Grifone
81 An algorithmic analysis of a communication model with retransmission of flawed messages
D M Lucantoni
82 Geometric games and their applications
W H Ruckle
83 Additive groups of rings
S Feigelstock
84 Nonlinear partial differential equations and their applications: Collège de France Seminar. Volume IV
H Brezis and J L Lions
85 Multiplicative functionals on topological algebras
T Husain
86 Hamilton-Jacobi equations in Hilbert spaces
V Barbu and G Da Prato
87 Harmonic maps with symmetry, harmonic morphisms and deformations of metrics
P Baird
88 Similarity solutions of nonlinear partial differential equations
L Dresner
89 Contributions to nonlinear partial differential equations
C Bardos, A Damlamian, J I Díaz and J Hernández
90 Banach and Hilbert spaces of vector-valued functions
J Burbea and P Masani
91 Control and observation of neutral systems
D Salamon
92 Banach bundles, Banach modules and automorphisms of C*-algebras
M J Dupré and R M Gillette
93 Nonlinear partial differential equations and their applications: Collège de France Seminar. Volume V
H Brezis and J L Lions
94 Computer algebra in applied mathematics: an introduction to MACSYMA
R H Rand
95 Advances in nonlinear waves. Volume I
L Debnath
96 FC-groups
M J Tomkinson
97 Topics in relaxation and ellipsoidal methods
M Akgül
98 Analogue of the group algebra for topological semigroups
H Dzinotyiweyi
99 Stochastic functional differential equations
S E A Mohammed
100 Optimal control of variational inequalities
V Barbu
101 Partial differential equations and dynamical systems
W E Fitzgibbon III
102 Approximation of Hilbert space operators. Volume II
C Apostol, L A Fialkow, D A Herrero and D Voiculescu
103 Nondiscrete induction and iterative processes
V Ptak and F-A Potra
104 Analytic functions – growth aspects
O P Juneja and G P Kapoor
105 Theory of Tikhonov regularization for Fredholm equations of the first kind
C W Groetsch

106 Nonlinear partial differential equations and free boundaries
J I Díaz

107 Variational convergences for functions and operators
H Attouch

108 A layering method for viscous, incompressible L_p flows occupying R^n
A Douglis and E B Fabes

M J Tomkinson

University of Glasgow

FC-groups

Pitman Advanced Publishing Program

BOSTON · LONDON · MELBOURNE

PITMAN PUBLISHING LIMITED
128 Long Acre, London WC2E 9AN

PITMAN PUBLISHING INC
1020 Plain Street, Marshfield, Massachusetts 02050

Associated Companies
Pitman Publishing Pty Ltd, Melbourne
Pitman Publishing New Zealand Ltd, Wellington
Copp Clark Pitman, Toronto

First published 1984

AMS Subject Classification: 20F24

Library of Congress Cataloging in Publication Data

Tomkinson, M. J.
FC-groups.

Bibliography: p.
Includes index.
1. FC-groups. I. Title. II. Title: FC-groups.
QA171.T65 1984 512′.2 83-26698
ISBN 0-273-08566-2

British Library Cataloguing in Publication Data

Tomkinson, M.J.
FC-groups.—(Research notes in mathematics; 96)
1. Finite groups
I. Title II. Series
512′.2 QA171

ISBN 0-273-08566-2

Reproduced and printed by photolithography
in Great Britain by Biddles Ltd, Guildford

Preface

One of the most influential papers in the study of FC-groups is that by Hall [49] in which he proved that a countable residually finite periodic FC-group can be embedded in a direct product of finite groups. Since then there have been a number of results which extend this theorem to certain uncountable groups. The most important of these is the result of Gorčakov that an FC-group which is a subgroup of a cartesian product of isomorphic finite groups can be embedded in a direct product of finite groups. Since Gorčakov's result appeared we have been able to obtain most of the previous results from his via a characterization of residually finite periodic FC-groups.

The main reason for writing these notes was to give a connected account of these results. This is given in Chapter 2 and the related Chapter 3 where we discuss the question of which periodic FC-groups are isomorphic to a section of a direct product of finite groups.

Having decided to give an account of these results it seemed a good opportunity to include other topics on FC-groups to illustrate the techniques which are now available in this area. These techniques have been developed specifically for FC-groups and tend to give the subject a character rather different from other areas of group theory. Indeed it is difficult to see how the results or techniques given here could be used in other areas, although I shall be delighted if people take up this challenge and prove me wrong.

Chapters 4 to 6 are concerned with inverse limit arguments, which are certainly the most widely used technique for proving theorems in FC-groups, it being possible to extend many results from finite groups to periodic FC-groups in this way. In Chapter 4 we consider profinite groups and by showing that the group of locally inner automorphisms of an FC-group is a profinite completion of the group of inner automorphisms are able to obtain most of the known results on locally inner automorphisms. The mostly well known results on Sylow theory in Chapter 5 illustrate how results can be built up from known results in the finite normal subgroups of G. We then use a combination of Sylow theory and inverse limit arguments applied to finite factor groups to develop a formation theory in locally soluble periodic FC-groups. It is possible to put these results on formations in a topological setting as with the results on locally inner automorphisms. However, at present there seems to be no gain in this approach and we prefer to keep the topological content to the minimum required for Chapter 4 where it undoubtedly shortens the proofs and gives a greater insight.

The results of Chapter 7 all derive from results of B.H. Neumann although some generalizations are given. The theme in this Chapter is the use of results from combinatorial set theory. Further generalizations of some results to $\mathfrak{m}$C-groups have appeared recently and these rely even more strongly on these combinatorial ideas. Finally, we include some topics which do not fit into any of the three main themes we have outlined but which we feel do form an important part of the subject.

One important topic which has been omitted is the study of specific bounds: for example, if $|G/Z(G)| = n$ then $|G'| \leqslant f(n)$; what is the function $f(n)$? The reason for omitting this type of question is that it reduces very quickly to a study of finite groups and more properly belongs in that area of group theory.

I have included some 25 unsolved questions. Although all such questions are important as showing some gap in our understanding, I believe that many of these are simply a matter of detail and can be answered using present techniques. However, I believe that some of these questions will require some totally new ideas. In particular, a number of questions in Chapter 3 require a much greater understanding of infinite extraspecial groups than we have at present. Perhaps even more important are Questions 4A and 5A. All the questions are phrased in such a way that I expect the answers to be positive - another challenge which I hope will be taken up.

The exercises at the end of the book are not exercises in the usual sense, rather they are a list of results which can be easily proved using the results and techniques contained here.

I have attempted to give a fairly comprehensive bibliography on FC-groups making considerable use of that given in Gorčakov's book [47] . Not all of the papers listed are referred to in the text. Some of these have been apparently superseded by later developments but do contain interesting techniques which may still prove to be of value.

The references in the text are less than comprehensive; it soon became clear that it was impossible to give references for proofs which are frequently an amalgam of different authors' methods. Some attempt has been made to give a reference to the first appearance of a theorem but even this has not always been possible and I apologise here for the errors and omissions which are sure to have occurred.

Although the subject matter of these notes is relatively self-contained we do of course need to refer to some text books. For infinite groups we refer to Robinson [84] , for finite groups Huppert [55] and for abelian groups Fuchs [35].

Finally I would like to record my gratitude to all those who have helped in the production of these Notes. A number of authors have sent me unpublished material which is referred to in the text. Brian Hartley and Stewart Stonehewer read parts of the manuscript and helped to remove a number of errors and obscurities; those that remain are, of course, in the parts that they did not read. I owe a special debt of gratitude to Stewart Stonehewer who, as my Ph.D. supervisor, first introduced me to the subject of FC-groups. Few research students have the good fortune to be directed into an area that will maintain their interest for the length of time that I have worked in this area. I would also like to thank Mrs. L. Williamson whose efficient typing of these Notes has done much to lessen this most tedious part of the author's task.

Glasgow
December 1983

M J Tomkinson

Contents

Notation

In the following, G is a group, S a subset of G, H and K subgroups of G, x and y elements of G.

$H \leq G$, $H < G$	H is a subgroup of G, H is a proper subgroup of G.
$H \lhd G$	H is a normal subgroup of G
$C_G(S)$	centralizer of S in G
$N_G(H)$	normalizer of H in G
$\langle S \rangle$	subgroup generated by S (also used for subspace generated by S in a vector space)
H^G	$\langle x^{-1}Hx : x \in G \rangle$, the normal closure of H in G
H_G	$\bigcap_{x \in G} x^{-1}Hx$, the core of H in G
$[x,y]$	$x^{-1}y^{-1}xy$
$[H,K]$	$\langle [h,k] : h \in H, k \in K \rangle$
G'	the derived subgroup of G
$Z(G)$	the centre of G
$Z_n(G)$	the nth term of the upper central series of G
$\mathrm{Dr}_{i \in I} G_i$	the (restricted) direct product of the groups $G_i, i \in I$.
$\Pi_{i \in I} G_i$	the cartesian product of the groups $G_i, i \in I$
$\mathrm{supp}(x)$	if $x \in \Pi_{i \in I} G_i$, $\mathrm{supp}(x) = \{i \in I$: ith component of x is non-trivial$\}$
$\mathrm{supp}(S)$	if $S \subseteq \Pi_{i \in I} G_i$, $\mathrm{supp}(S) = \bigcup_{x \in S} \mathrm{supp}(x)$
$\mathfrak{m}$	an infinite cardinal
$\mathfrak{m}^+$	the cardinal successor to $\mathfrak{m}$
$\exp \mathfrak{m}$	the cardinal of the power set of $\mathfrak{m}$

[We usually distinguish between cardinals and ordinals and use Greek letters to denote ordinals.]

Mappings may be written on either the right or left.

The end or absence of a proof is indicated by the symbol □.

1 Basic results

If x is an element of the group G, then we use $C\ell(x)$ (or $C\ell_G(x)$) to denote the conjugacy class of G containing x; that is, the set of all elements of G which are conjugate to x. The group G is said to be an FC-group (or G has finite classes) if $C\ell(x)$ is finite for each $x \in G$. This is, of course, equivalent to saying that $|G:C_G(x)|$ is finite for each $x \in G$.

It is clear that all finite groups and all abelian groups are FC-groups and most of the work on FC-groups can be interpreted as extending known results for one or other of these classes. Further examples of FC-groups can be obtained by noting that the class of FC-groups is closed under the operations Q (taking homomorphic images), S (taking subgroups) and D (forming restricted direct products). Thus, if, for each $i \in I$, G_i is a finite group, G is the direct product $G = \mathrm{Dr}_{i \in I}\, G_i$, H is a subgroup of G and K is a normal subgroup of H, then H/K is an FC-group since it is in the class $QSD\mathfrak{F}$ (where $\mathfrak{F}$ denotes the class of finite groups). One of the more interesting questions concerning FC-groups is that of determining when an FC-group is in this class $QSD\mathfrak{F}$ (see Chapter 3).

The condition of being an FC-group is, of course, a very strong condition and the class of FC-groups is not closed under any of the other commonly used closure operations. For example, let C be the extension of a countably infinite elementary abelian 3-group A by a group B of order 2 in which the generator of B inverts each element of A. Then G is not an FC-group although it is locally finite and residually finite and is a finite extension of an abelian group. If either of the groups H,K are infinite, then the wreath product H wr K will contain elements with infinitely many conjugates.

However we can combine finite groups and abelian groups in other ways to obtain FC-groups.

<u>Theorem 1.1</u> <u>If</u> G <u>is a group with finite derived group</u> G', <u>then</u> G <u>is an FC-group</u>.

Proof. If $x \in G$, then every conjugate of x is contained in the finite set xG', since $g^{-1}xg = x[x,g]$. □

Theorem 1.2. (Schur [86]) *If G is a group in which the centre Z has finite index, then G' is finite.*

Proof. Let $\{t_1, \ldots, t_n\}$ be a transversal to Z in G. Then each commutator of G is of the form $[t_i, t_j]$ and so there are at most $n(n-1)$ distinct commutators. The elements of G' are finite products of commutators and we show that each element of G' can be expressed as a product of at most $n(n-1)^2$ commutators.

Let $x \in G'$ and suppose that x can be expressed as a product of r commutators but can not be written as a product of fewer commutators. If $r > n(n-1)^2$ then one of the commutators must appear in the product n times. These n commutators can be brought to the front of the product and the remaining r-n commutators will be replaced by conjugates so that we can write

$$x = [a,b]^n [a_{n+1}, b_{n+1}] \ldots [a_r, b_r] .$$

It is easily verified that

$$[a,b]^n = (a^{-1}b^{-1})^n (ab)^n c_{n-1} c_{n-2} \ldots c_1,$$

where c_k is the commutator $[(ab)^{-(k+1)} (a^{-1}b^{-1})^k (ab)^{(k+1)}, (ab)^{-1}]$. But $(ab)^n \in Z$ and so $(ab)^n = b(ab)^n b^{-1} = (ba)^n$, the inverse of $(a^{-1}b^{-1})^n$. Hence $[a,b]^n = c_{n-1} c_{n-2} \ldots c_1$ can be written as a product of n-1 commutators and hence x can be written as a product of r-1 commutators, contrary to the minimality of r. It follows that $r \leq n(n-1)^2$. Hence G' is finite and the order of G' is bounded by a function of n. □

The proof given above is attributed to D. Ornstein in [56]. There are a number of other, shorter, proofs but these all make use of the transfer map.

We have shown that centre-by-finite groups are finite-by-abelian and these in turn are FC-groups. Generally speaking, from our point of view, centre-by-finite groups are very simple objects and the characterizations given in Chapter 7, for example, really say that groups with certain properties have a rather simple structure. However, we shall see that finite-by-abelian groups may have rather unusual properties and we shall frequently consider particular examples of groups of this type.

The results which follow will show that in many situations it is natural to consider periodic FC-groups. These groups are also known (particularly in Russian articles) as locally normal groups or occasionally as locally finite-normal groups. The reason for this terminology is seen in the following important and fundamental result usually referred to as Dicman's Lemma. We base our proof on the theorem given above although the results can be proved directly (see, for example, [84] Part 1, p.45).

Lemma 1.3. (Dicman [19]) *Let* $x_1,\ldots,x_n$ *be elements of the group* G, *each having finite order and each having only finitely many conjugates in* G. *Then there is a finite normal subgroup* N *of* G *containing* $x_1,\ldots,x_n$.

Proof. For each $i=1,\ldots,n$, let $C_i = C_G(x_i)$ and let $K_i = (C_i)_G$, the core of C_i. Then G/K_i is finite and if $K = K_1\cap\ldots\cap K_n$, then G/K is finite. Now K centralizes each conjugate of each x_i and so, if $N = \langle x_1,\ldots,x_n\rangle^G$ is the subgroup generated by all the conjugates of $x_1,\ldots,x_n$, then K centralizes N. Hence $K\cap N \leq Z(N)$ and so $N/Z(N)$ is finite. By Theorem 1.2, N' is finite. Also N is finitely generated and so the finitely generated abelian group N/N' has finite torsion subgroup T/N'. Since N/T is torsion-free, each conjugate of each x_i must belong to T and hence N = T and so N is finite. □

In any group G, it is clear that the elements x such that $C\ell(x)$ is finite form a subgroup $\Delta(G)$ and this subgroup is characteristic in G. One consequence of Dicman's Lemma is that the elements of $\Delta(G)$ of finite order also form a characteristic subgroup $\Delta^+(G)$. Any finite set of elements of $\Delta^+(G)$ is contained in a finite normal subgroup of G and in particular $\Delta^+(G)$ is locally finite. If G is a periodic FC-group then $\Delta^+(G) = G$ and so each finite set of elements of G is contained in a finite normal subgroup. It is clear that a group satisfying this locally finite-normal condition must be a periodic FC-group.

The next few results give some justification for our concentrating on periodic FC-groups. Essentially they say that if G is an FC-group then it becomes periodic if we factor out a "nice" subgroup (the centre) or if we remove a "nice" factor group (the derived factor group).

Theorem 1.4. (Baer [4]) *If G is an FC-group, then G/Z(G) is periodic.*

Proof. Let $x \in G$ and let $C = C_G(x)$ so that $|G:C|$ is finite. Let $\{t_1, \ldots, t_n\}$ be a (right) transversal to C in G and consider $D = \bigcap_{i=1}^{n} C_G(t_i)$. Then $|G:D|$ is finite and hence $x^k \in D$ for some positive integer k. Then x^k centralizes $\{t_1, \ldots, t_n\}$ and also centralizes C; thus $x^k \in Z(G)$, as required.□

If G is a finitely generated FC-group then G/Z(G) is a finitely generated periodic FC-group. It follows from Dicman's Lemma 1.3 that G/Z(G) is finite and hence Z(G) is finitely generated. We therefore have the following result.

Corollary 1.5. (Neumann [70]) (i) *A group G is a finitely generated FC-group if and only if G is a finite extension of a central finitely generated abelian group.*

(ii) *A subgroup of a finitely generated FC-group is finitely generated. Hence an FC-group satisfies the maximal condition if and only if it is finitely generated.* □

Theorem 1.6. (Neumann [70]) *If G is an FC-group then G' is periodic and hence the elements of finite order form a subgroup of G containing G'.*

Proof. Each element x of G' is a product of finitely many commutators and so there is a finitely generated subgroup H of G such that $x \in H'$. By Corollary 1.5, H is centre-by-finite and so, by Theorem 1.2, H' is finite. Hence x has finite order. □

Theorem 1.7. (Černikov [15]) (i) *A group G is an FC-group if and only if it is isomorphic to a subgroup of the direct product of a periodic FC-group and a torsion-free abelian group.*

(ii) *A group G is a finitely generated FC-group if and only if it is isomorphic to a subgroup of the direct product of a finite group and a free* abelian group of finite rank.

Proof. The "if" direction in each of these results is clear from the closure properties of the class of FC-groups stated in our opening remarks.

So let G be an FC-group. Let T be the subgroup consisting of all elements of finite order (Theorem 1.6) so that $T \geq G'$ and G/T is a torsion-free abelian group. If G is finitely generated then so is G/T and hence G/T is a free abelian group of finite rank. The centre Z of G contains a maximal torsion-free subgroup A so that G/A is periodic and if G is finitely generated then G/A will be finite. Clearly $T \cap A = 1$ and so there is a natural embedding of G into $(G/A) \times (G/T)$ mapping x to (xA,xT). This completes the proof of the theorem. □

This last result can be interpreted as saying that if we know all periodic FC-groups (and all torsion-free abelian groups) then we can construct all FC-groups. Again this gives some justification for our concentration on periodic FC-groups although we should say that this last theorem does not give an entirely satisfactory classification of FC-groups as it is extremely difficult to determine all subgroups of a direct product. There are also some interesting questions concerning nonperiodic FC-groups which can not be answered by simply referring to the periodic case (see, for example, p. 40).

The results above have used Dicman's Lemma to concentrate attention on either a periodic subgroup or periodic factor group of an FC-group. When G is a periodic FC-group then it is generated by its finite normal subgroups but often we do not need to consider all finite normal subgroups but only a local system in the following sense.

A *local system* of normal subgroups of G is a family $\{F_i : i \in I\}$ of normal subgroups of G such that

(LS1) $G = \bigcup_{i \in I} F_i$,

(LS2) for each $i,j \in I$, there is a $k \in I$ such that $F_i F_j \leq F_k$.

We shall be concerned with local systems of finitely generated normal subgroups or, in the periodic case, with local systems of finite normal subgroups. Since the set of all finite normal subgroups of a periodic FC-group clearly forms a local system, we have

Theorem 1.8. A group G is a periodic FC-group if and only if G has a local system of finite normal subgroups. □

This will clearly be of great importance in what follows. But the finite factor groups will also play an important role. Since $Z(G) = \bigcap_{x\in G} C_G(x)$ and $|G:C_G(x)|$ is finite, we have

Theorem 1.9. If G is an FC-group, then G/Z(G) is residually finite. Hence any subgroup H of G which has no proper subgroup of finite index is a divisible abelian group and is contained in Z(G). □

Again we do not always need to consider all normal subgroups of finite index and dualizing the concept of a local system we define a *residual system* of normal subgroups of finite index in G to be a family $\{N_i : i\in I\}$ of normal subgroups of finite index in G such that

(RS1) $1 = \bigcap_{i\in I} N_i$,

(RS2) for each i, j∈I, there is a k∈I such that $N_i N_j \geq N_k$.

Corresponding to Theorem 1.8, we have

Theorem 1.10. A group G is residually finite if and only if G has a residual system of normal subgroups of finite index. □

Although it is possible to consider residual systems of normal subgroups such that the factor groups satisfy any required condition we shall only consider those defined above. We shall therefore frequently refer just to a residual system and it will be understood that the normal subgroups have finite index.

One immediate consequence of a group G being residually finite is that G can be embedded in a cartesian product of finite groups. Specifically, if $\{N_i : i\in I\}$ is a residual system then the mapping $x \mapsto (xN_i)_{i\in I}$ embeds G in the cartesian product $\prod_{i\in I} (G/N_i)$. This type of embedding is considered in much more detail in Chapter 2.

Theorems 1.4 and 1.9 combine to show that if G is an FC-group then $G/Z(G)$ is a residually finite periodic FC-group and we will often be able to restrict our attention to this class of groups. One would expect residual finiteness to be less useful than the very strong locally normal condition but, in fact, when working with both conditions together we often seem to make more use of the residual finiteness. For example, the group Inn G of inner automorphisms of G is isomorphic to $G/Z(G)$ and we shall see that the residual finiteness of that group plays a crucial role in Chapter 4. In that chapter we shall need to consider different residual systems of $G/Z(G)$ and we indicate below how these are obtained.

Lemma 1.11. *Let Z be the centre of the FC-group G.*

(i) *Let* $\{F_i : i\in I\}$ *be a local system of finitely generated subgroups of G. Then the subgroups* $\{C_G(F_i)/Z\}$ *form a residual system of* G/Z.

(ii) *Let G be residually finite and let* $\{N_i : i\in I\}$ *be a residual system of G. For each* $i\in I$, *let* $C_i = C_G(G/N_i)$; *then the families* $\{C_i/Z : i\in I\}$ *and* $\{ZN_i/Z : i\in I\}$ *are both residual systems of* G/Z.

Proof. (i) Since G is an FC-group, $C_G(F_i)$ has finite index in G and clearly $\bigcap_{i\in I} C_G(F_i) = Z(G)$. If $i,j\in I$ then there is a $k\in I$ such that $F_iF_j \leq F_k$ and hence $C_G(F_k) \leq C_G(F_i)C_G(F_j)$.

(ii) Clearly G/ZN_i is finite and, if $k\in I$ is such that $N_i \cap N_j \geq N_k$, then we also have $ZN_i \cap ZN_j \geq ZN_k$. We have only to show that $\bigcap_{i\in I} ZN_i = Z$ and this follows from $[G,\bigcap_{i\in I} ZN_i] \leq \bigcap_{i\in I} [G,ZN_i] \leq \bigcap_{i\in I} N_i = 1$.

Similarly G/C_i is finite and if $k\in I$ is such that $N_i \cap N_j \geq N_k$ then C_k centralizes both G/N_i and G/N_j so that $C_i \cap C_j \geq C_k$. Also $[G,\bigcap_{i\in I} C_i] \leq \bigcap_{i\in I} [G,C_i] \leq \bigcap_{i\in I} N_i = 1$ and so $\bigcap_{i\in I} C_i = Z$. □

The fact that a periodic FC-group has a local system of finite normal subgroups gives us a great deal of information about series in G and the next few results illustrate ways in which this can be obtained. We use the terminology of Robinson's book [84] for series but we shall see that we usually only need to consider ascending or descending series.

Theorem 1.12. (Tôgô [97]) Let H be a subgroup of the periodic FC-group G and let $\{F_i : i \in I\}$ be a local system of finite normal subgroups of G. Then the following are equivalent:

(a) H is a serial subgroup of G,

(b) $H \cap F_i$ is a subnormal subgroup of F_i, for each $i \in I$,

(c) $H \triangleleft^{\omega} G$; that is, there is an ascending series of type ω from H to G.

Proof. It is immediate that (a) implies (b) and that (c) implies (a) so that we need only prove that (b) $\Rightarrow$ (c).

We show first that if $H \cap F_i < F_i$ then F_i has a subnormal subgroup K_i such that $H \cap F_i < K_i$ and $K_i \leq N_G(H)$. If this is not the case, then for each subnormal subgroup K of F_i such that $H \cap F_i < K$ there is a $j = j(K) \in I$ such that K does not normalize $F_j \cap H$. Thus K does not normalize $F_k \cap H$ for any $k \in I$ such that $F_j \leq F_k$. Since there are only finitely many such subgroups K, there is a $k \in I$ such that F_k contains each F_j and also $F_k \geq F_i$. It follows that no subnormal subgroup K of F_i such that $H \cap F_i < K$ normalizes $F_k \cap H$. But there is a series

$$F_k \cap H = A_o \triangleleft A_1 \triangleleft \ldots \triangleleft A_r = F_k.$$

Let s be minimal such that $F_i \cap A_s > F_i \cap H$. Then $F_i \cap A_s$ is subnormal in F_i and $[F_i \cap A_s, F_k \cap H] \leq [F_i \cap A_s, A_{s-1}] \leq F_i \cap A_{s-1} = F_i \cap H \leq F_k \cap H$ so that $F_i \cap A_s$ normalizes $F_k \cap H$. This contradiction proves the existence of a subnormal group K of F_i such that $F_i \cap H < K$ and $K \leq N_G(H)$.

Since F_i is finite, the join X_i of all subnormal subgroups of F_i which normalize H is also a subnormal subgroup of F_i and, by the above, $H \cap F_i < X_i$ whenever $H \cap F_i < F_i$. If $F_i \leq F_j$, then since X_i is a subnormal subgroup of F_j normalizing H, we have $X_i \leq X_j$. Also $X_j \cap F_i$ is a subnormal subgroup of F_j normalizing H and so $X_j \cap F_i = X_i$. If we define $H_1 = \bigcup_{i \in I} X_i$, then H_1 normalizes H and $H_1 \cap F_i = X_i$ is a subnormal subgroup of F_i which properly contains $H \cap F_i$ whenever $H \cap F_i < F_i$.

Now $H_1 \cap F_i$ is a subnormal subgroup of F_i for each $i \in I$ and so we can construct a subgroup H_2 such that $H_1 \triangleleft H_2$ and, whenever $H_1 \cap F_i < F_i$,

$H_2 \cap F_i$ properly contains $H_1 \cap F_i$. Continuing in this way we obtain an ascending series

$$H \lhd H_1 \lhd H_2 \lhd \ldots$$

and, since each F_i is finite, there is a positive integer $n = n(i)$ such that $F_i \cap H_n = F_i$ so that $F_i \leq H_n$ and $G = \bigcup_{n=1}^{\infty} H_n$. □

This theorem can be easily applied in the nonperiodic case by noting that H is a serial subgroup of G if and only if HZ/Z is a serial subgroup of G/Z.

Theorem 1.13. (i) (a) A simple FC-group is finite and hence every composition factor of an FC-group is finite.

(b) A minimal normal subgroup of an FC-group is finite and hence each chief factor of an FC-group is finite.

(ii) An infinite periodic FC-group G has an ascending composition series (chief series) $\mathcal{C}$ of type ρ, where ρ is the least ordinal of cardinality $|G|$. Each composition factor (chief factor) of G is isomorphic to a factor in $\mathcal{C}$.

Proof. (i) A minimal normal subgroup of an FC-group is finitely generated and so the result follows from Corollary 1.5 (i).

(ii) The elements of G may be well-ordered so that $G = \{x_\alpha : \alpha < \rho\}$. For each $\alpha < \rho$, let F_α be a finite normal subgroup of G containing x_α. If $H_\alpha = \langle F_\beta : \beta < \alpha \rangle$ then the $H_\alpha, \alpha < \rho$, form an ascending normal series of G with finite factors $H_{\alpha+1}/H_\alpha$. This can clearly be refined to an ascending composition series (chief series) of G of type ρ by inserting finitely many subgroups between each pair H_α and $H_{\alpha+1}$.

If U/V is any composition factor (chief factor) of G, then U/V is finite. Therefore, there is a finite normal subgroup F of G such that $FV \geq U$ and so $U/V \cong (F \cap U)/(F \cap V)$. If $\mathcal{C} = (G_\alpha : \alpha < \rho)$, then the subgroups $F \cap G_\alpha$ form a composition series (G-composition series) of F when duplications are omitted and so there is a factor $(F \cap G_{\alpha+1})/(F \cap G_\alpha)$ which is isomorphic to $(F \cap U)/(F \cap V)$. But then

$$U/V \cong (F \cap U)/(F \cap V) \cong (F \cap G_{\alpha+1})/(F \cap G_\alpha) \cong G_{\alpha+1}/G_\alpha,$$

as required. □

The second part of the above result does, of course, require G to be periodic as, for example, the infinite cyclic group does not have an ascending composition series.

Further useful series may be obtained by considering the *socle* of G. If G is any group then Soc(G) is the subgroup generated by all minimal normal subgroups of G; if G is a periodic FC-group then Soc(G) is a direct product of finite normal subgroups of G. The *socle series* of a group G is defined by taking $S_o = 1$ and $S_{\alpha+1}/S_\alpha = Soc(G/S_\alpha)$. If α is a limit ordinal then one defines $S_\alpha = \bigcup_{\beta<\alpha} S_\beta$ but this is not needed for periodic FC-groups.

Lemma 1.14. *If S_n are terms in the socle series of the periodic FC-group G, then $\bigcup_{n=1}^{\infty} S_n = G$.*

Proof. A periodic FC-group G is the union of its finite normal subgroups F_i. Suppose that F_i has a G-composition series

$$1 = G_o < G_1 < \ldots < G_n = F_i ,$$

and assume inductively that $G_{k-1} \leqslant S_{k-1}$. Then either $G_k \leqslant S_{k-1}$ or $G_k S_{k-1}/S_{k-1}$ being G-isomorphic to G_k/G_{k-1} is a minimal normal subgroup of G/S_{k-1}. In both cases we obtain $G_k \leqslant S_k$. Hence $F_i \leqslant S_n$ and so $G = \bigcup_{n=1}^{\infty} S_n$. □

Corollary 1.15. *A locally soluble (locally nilpotent) FC-group has an ascending abelian normal series (central series) of type* ω.

Proof. Let the periodic FC-group G/Z have socle series

$$1 = S_o/Z \leqslant S_1/Z \leqslant \ldots \leqslant \bigcup_{n=1}^{\infty} S_n/Z = G/Z ;$$

then, in both cases, the required series is

$$1 \leqslant Z \leqslant S_1 \leqslant \ldots \leqslant \bigcup_{n=1}^{\infty} S_n = G. \ \square$$

This result says in particular that locally soluble FC-groups are hyperabelian (or SI* -groups in Kuroš' notation) and locally nilpotent

FC-groups are hypercentral (or ZA-groups).

Descending series may be obtained using the residual finiteness of G/Z.

Theorem 1.16. (Stonehewer [93]) A locally soluble (locally nilpotent) FC-group has a descending abelian normal series (central series) of type $\omega+1$.

Proof. By considering G/Z, it is sufficient to show that a residually finite locally soluble (locally nilpotent) group G has a descending abelian (central) series of type ω. So let $\{N_i : i \in I\}$ be a residual system of normal subgroups of finite index in G. Then G/N_i is a finite soluble (nilpotent) group. Let G_n be the nth term of the derived series (lower central series) of G, so that, for each $i \in I$, there is an integer n(i) such that $N_i \geqslant G_{n(i)}$. Then $\bigcap_{n=1}^{\infty} G_n \leqslant \bigcap_{i \in I} N_i = 1$, as required.□

This result can not be improved to give a descending series of type ω even in the periodic case. For, let X_n be a finite p-group of derived length n and let Z_n be a central subgroup of order p contained in $X_n^{(n-1)}$, the (n-1)th term of the derived series of X_n. Form G, the direct product of the groups X_n with the subgroups Z_n amalgamated; then the derived series of G has type $\omega+1$.

Further results on chief factors and chief series can be obtained from known results on finite groups. If π is a set of primes, then a π-group is a periodic group in which the order of each element is a product of primes from the set π. It is clear that the product of all the normal π-subgroups of a group G is also a normal π-subgroup. This is the unique largest normal π-subgroup of G called the π-*radical* of G and denoted by $O_\pi(G)$. Of particular interest are the subgroups $O_{p'}(G)$, where p′ is the set of all primes other than p, and $O_{p'p}(G)$ defined by $O_{p'p}(G)/O_{p'}(G) = O_p(G/O_{p'}(G))$. A finite group G is said to be p-*nilpotent* if $O_{p'p}(G) = G$. A periodic FC-group is *locally* p-*nilpotent* if $O_{p'p}(G) = G$. We also define the *locally nilpotent radical* R(G) of a group G to be the unique largest normal locally nilpotent subgroup of G. If G is finite then R(G) is the *Fitting subgroup* of G. In a periodic FC-group these radicals are easily obtainable via a local system of finite normal subgroups of G or a residual system of G/Z.

Lemma 1.17. *Let G be a periodic FC-group.*

(i) *Let* $\{F_i: i \in I\}$ *be a local system of finite normal subgroups of G. Then* $O_{p'p}(G) = \bigcup_{i\in I} O_{p'p}(F_i)$ *and* $R(G) = \bigcup_{i\in I} R(F_i)$.

(ii) *Let* $\{N_i/Z: i \in I\}$ *be a residual system of normal subgroups of finite index in G/Z. Then* $O_{p'p}(G) = \bigcap_{i\in I} M_i$, *where* $M_i/N_i = O_{p'p}(G/N_i)$, *and* $R(G) = \bigcap_{i\in I} R_i$, *where* $R_i/N_i = R(G/N_i)$.

Proof. (i) Since $O_{p'p}(F_i)$ is characteristic in F_i, it is normal in G and hence $O_{p'p}(F_i) \leq O_{p'p}(G)$. Conversely, $F_i \cap O_{p'p}(G)$ is a normal p-nilpotent subgroup of F_i and so is contained in $O_{p'p}(F_i)$. It follows that $O_{p'p}(G) = \bigcup_{i\in I} (F_i \cap O_{p'p}(G)) \leq \bigcup_{i\in I} O_{p'p}(F_i)$. The proof for R(G) is similar.

(iii) It is clear that $O_{p'p}(G) \leq M_i$ and so we must prove that $\bigcap_{i\in I} M_i$ is locally p-nilpotent. Let F be a finite normal subgroup of G contained in $\bigcap_{i\in I} M_i$. There is an $i \in I$ such that $N_i \cap F \leq Z$ and, since $F/(F \cap N_i) \cong FN_i/N_i \leq M_i/N_i$, we see that $F/(F \cap Z)$ is p-nilpotent. Since $F \cap Z \leq Z(F)$ it follows that F is p-nilpotent, as required. Again the proof for R(G) is similar. □

Theorem 1.18. *Let* $\mathcal{C}$ *be any chief series of the periodic FC-group G. Then*

(i) $O_{p'p}(G) = \bigcap\{C_G(U/V): U/V$ *a chief factor of G such that* $p \mid |U/V|\}$

$= \bigcap\{C_G(U/V): U/V$ *a factor in* $\mathcal{C}$ *such that* $p \mid |U/V|\}$.

(ii) $R(G) = \bigcap\{C_G(U/V): U/V$ *a chief factor of G* $\}$

$= \bigcap\{C_G(U/V): U/V$ *a factor in* $\mathcal{C}$ $\}$.

Proof. These results are well known for finite groups ([55] , pp.278, 686) and can be extended to G by using Lemma 1.17.

Let $\{N_i/Z: i \in I\}$ be the set of all normal subgroups of finite index in G/Z and, for each $i \in I$, let $\mathcal{C}_i$ be the chief series of G/N_i obtained by removing repetitions from the subgroups CN_i, $C \in \mathcal{C}$. Then the result for finite groups says that

$$M_i = \bigcap\{C_G(U_i/V_i): U_i/V_i \text{ a factor in } \mathcal{C}_i \text{ such that } p \mid |U_i/V_i|\}.$$

Since $U_i/V_i = UN_i/VN_i \cong U/V$ for some factor U/V in $\mathcal{C}$, we have

$$O_{p'p}(G) = \bigcap_{i \in I} M_i \geq \bigcap\{C_G(U/V): U/V \text{ a factor in } \mathcal{C} \text{ such that } p \mid |U/V|\}.$$

Now let U/V be a factor in $\mathcal{C}$ with $p \mid |U/V|$ and let $Y/V = Z(G/V)$ so that $Z \leq Y$. If $U \leq Y$ then $C_G(U/V) = G$. If $U \nleq Y$ then since G/Y is residually finite there is an $i \in I$ such that $N_i \cap U = V$ and so $U/V \cong UN_i/VN_i$. Therefore each non-central factor in $\mathcal{C}$ is G-isomorphic to a factor in some $\mathcal{C}_i$ and so

$$O_{p'p}(G) = \bigcap_{i \in I} M_i = \bigcap\{C_G(U/V): U/V \text{ a factor in } \mathcal{C} \text{ such that } p \mid |U/V|\}.$$

The other parts of the theorem may be proved by similar arguments. □

In one of the results above we used the fact that if U/V is a finite factor in a periodic FC-group G then there is a finite normal subgroup F of G such that $FV \geq U$. This extends easily to infinite factors.

Lemma 1.19. *Let G be an FC-group and let $V \triangleleft U \leq G$. If U/V is infinite then there is a normal subgroup F of G such that $|F| = |U/V|$ and $FV \geq U$.*

Proof. Let $\{u_i : i \in I\}$ be a transversal to V in U and let $F = \langle u_i : i \in I \rangle^G$. □

Corollary 1.20. *Let G be a locally soluble FC-group and let $\mathfrak{m}$ be an infinite cardinal. If $|G/G'| < \mathfrak{m}$, then $|G| < \mathfrak{m}$.*

Proof. If G/G′ is infinite then there is a normal subgroup F such that $|F| = |G/G'|$ and $FG' = G$. Therefore G/F is perfect, but also G/F is locally soluble and so, by Theorem 1.16, $F = G$ and hence $|G| = |G/G'|$.

If G/G′ is finite then there is a finitely generated normal subgroup F such that $FG' = G$. As above, we can show that $F = G$ and so G is finitely generated. But a finitely generated FC-group has a finite derived subgroup G′ (Corollary 1.5 and Theorem 1.2) and so G is finite. □

2 Subgroups of direct products of finite groups

In Chapter 1 we observed that a periodic FC-group is generated by its finite normal subgroups and that G/Z(G) is residually finite so that, in some sense, G is built up from its finite normal subgroups and a large part of G is similarly dependent on its finite factor groups. In Chapters 5 and 6 we shall see how results from the theory of finite groups can frequently be extended to periodic FC-groups by building up the result through the finite normal subgroups or finite factor groups.

In this and the following Chapter we are more concerned with the question of how periodic FC-groups can actually be constructed from finite groups. Any (restricted) direct product of finite groups is clearly a residually finite periodic FC-group and so too is any subgroup. That is, the class $SD\mathfrak{F}$ is contained in the class of residually finite periodic FC-groups. If we take a homomorphic image of such a group then we may lose the residual finiteness but will still obtain a periodic FC-group. That is, the class $QSD\mathfrak{F}$ is contained in the class of periodic FC-groups.

In an important paper in 1959, P. Hall [49] showed that the converses of these two results are true for countable groups. That is, all countable (residually finite) periodic FC-groups can be constructed in the above manner. Unfortunately, neither result extends fully to uncountable groups but we shall see in these two Chapters that the classes $QSD\mathfrak{F}$ and $SD\mathfrak{F}$ are very large subclasses of the classes of periodic FC-groups and residually finite periodic FC-groups.

We deal in this Chapter with residually finite periodic FC-groups and in particular with the question of when such a group can be embedded in a direct product of finite groups. This question may be compared with the problem of determining which periodic abelian groups without elements of infinite height are isomorphic to a direct product of finite cyclic groups or, equivalently, to a direct product of finite abelian groups. One important difference is that every subgroup of a direct product of finite abelian groups is itself a direct product of finite groups ([35] , p.91) but if one omits the abelian condition this is no longer true.

Example 2.1. (Hall [49]) For each integer n, let G_n be a dihedral group of order 8, say $G_n = \langle a_n, b_n : a_n^2 = b_n^2 = 1, [a_n, b_n] = c_n$, where c_n is a central element of order $2\rangle$. Let G be the subgroup of $\mathrm{Dr}_{n=-\infty}^{\infty} G_n$ generated by the elements $g_{2n-1} = a_{2n-1} a_{2n}$ and $g_{2n} = b_{2n} b_{2n+1}$. Note that $[g_{2n-1}, g_{2n}] = c_{2n}$ and $[g_{2n}, g_{2n+1}] = c_{2n+1}$ so that $G' = Z(G) = \mathrm{Dr}_{n=-\infty}^{\infty} \langle c_n \rangle$. Each non-central element of G can be written uniquely as a product

$$g = g_{i(1)} \cdots g_{i(r)} z$$

with $i(1) < \ldots < i(r)$ and $z \in G'$. We shall say that $g_{i(1)}$ is the initial component and $g_{i(r)}$ the final component of g.

Suppose that $G = H \times K$ and that h,k are non-central elements of H,K respectively. If h and k had the same initial component g_i, then we would have $[g_{i-1}, h] = [g_{i-1}, k] = c_i \neq 1$ contrary to $[g_{i-1}, h] \in H$, $[g_{i-1}, k] \in K$ and $H \cap K = 1$. Thus h and k have different initial components. Similarly, if they had the same final component g_j, then we would have $[h, g_{j+1}] = [k, g_{j+1}] = c_{j+1} \neq 1$ again giving a contradiction. It follows that when hk is expressed in the standard form above we must have $r \geqslant 2$. Thus the element g_n is not a product of non-central elements h,k and so g_n must belong to HG' or to KG'. Therefore, for each integer n, there is a $z_n \in G'$ such that $g_n z_n \in H$ or K. But $[g_n z_n, g_{n+1} z_{n+1}] = [g_n, g_{n+1}] = c_{n+1} \neq 1$ and so $g_n z_n$ and $g_{n+1} z_{n+1}$ belong to the same factor. Hence each g_n is contained in HZ (or KZ) and so $G = HZ$ (or $G = KZ$). Also $c_n = [g_{n-1}, g_n] \in H$ and so $Z \leqslant H$ giving $G = H$. Therefore G has no direct decomposition. □

Despite this complication subgroups of direct products of finite groups are easier to deal with than arbitrary residually finite periodic FC-groups. We show in our first result that the factors in the direct product can be ordered in a particularly convenient way and this will enable us to obtain results about the derived subgroup G' and the central factor group $G/Z(G)$ of an $\mathcal{SDF}$-group. Before stating this theorem we describe our notations for projections in cartesian or direct products. Let I be an index set and J a subset of I; then π_J will denote the projection map from the cartesian product $\Pi_{i \in I} F_i$ to $\Pi_{i \in J} F_i$. Contrary to our normal practice we write these projection maps on the left so that, for example $\pi_j((g_i)_{i \in I}) = g_j$. We also use a second notation for projections: if G is a subgroup of

$\Pi_{i\in I} F_i$ or if g is an element of $\Pi_{i\in I} F_i$ then we sometimes denote $\pi_J(G)$ by $G(J)$ or $\pi_J(g)$ by $g(J)$. This second notation is not strictly necessary but is rather more compact particularly when the form of the subset J is complicated. We prefer to retain the more usual $\pi_J(G)$ notation if the subgroup G is expressed in a complicated form as, for example, in the statement of the next result.

Theorem 2.2. (Gorčakov [47]) Let G be a subgroup of the direct product of the finite groups $F_i, i\in I$. Then the index set I may be expressed as a union of an ascending chain of sets $I(\alpha), \alpha < \rho$, where ρ is the least ordinal of cardinality $|I|$, such that

(i) $J(\alpha) = I(\alpha+1) - I(\alpha)$ is a countable set,

(ii) for each finite subset J of $I(\alpha)$,
$\pi_J(G) = \pi_J(G \cap \mathrm{Dr}_{i\in I(\alpha)} F_i)$.

Proof. We may assume that the index set is well-ordered and so $I = \{\alpha : \alpha < \rho\}$. We construct the sets $I(\alpha)$ inductively to satisfy conditions (i)and (ii) above and also $\alpha \in I(\alpha+1)$ so that $\bigcup_{\alpha<\rho} I(\alpha) = I$. The construction can be started by taking $I(1) = \emptyset$ and we suppose therefore that the sets $I(\beta)$ have been constructed for all $\beta < \alpha$.

If α is a limit ordinal, then we define $I(\alpha) = \bigcup_{\beta<\alpha} I(\beta)$. If J is a finite subset of $I(\alpha)$ then there is a $\gamma = \gamma(J) < \alpha$ such that $J \subseteq I(\gamma)$. Therefore

$$\pi_J(G) = \pi_J(G \cap \mathrm{Dr}_{i\in I(\gamma)} F_i) \leqslant \pi_J(G \cap \mathrm{Dr}_{i\in I(\alpha)} F_i) \leqslant \pi_J(G).$$

Suppose then that $\alpha = \beta+1$. If $\beta \in I(\beta)$ then we can define $I(\alpha) = I(\beta)$. If $\beta \notin I(\beta)$, let $J_o = \{\beta\}$. There is a finite subgroup X_1 of G such that $X_1(J_o) = G(J_o)$. Let J_1 be the finite set $J_o \cup (\mathrm{supp}\,(X_1) - I(\beta))$. There is a finite subgroup X_2 of G such that $X_2(J_1) = G(J_1)$. We now let J_2 be the finite set $J_1 \cup (\mathrm{supp}(X_2) - I(\beta))$. Continuing in this way we can define an ascending chain of finite subsets J_n of $I - I(\beta)$ and a sequence of finite subgroups X_n of G such that $X_n(J_{n-1}) = G(J_{n-1})$ and $J_n = J_{n-1} \cup (\mathrm{supp}(X_n) - I(\beta))$.

Define $J(\beta) = \bigcup_{n=0}^{\infty} J_n$; then $J(\beta)$ is countable and $I(\beta) \cap J(\beta) = \emptyset$. Define $I(\alpha) = I(\beta) \cup J(\beta)$. If J is any finite subset of $I(\alpha)$, then

$J \subseteq (J \cap I(\beta)) \cup J_n$, for some n. Let $g \in G$; then there is an element $x \in X_{n+1}$ such that $x(J_n) = g(J_n)$. By the induction hypothesis, there is an element $y \in G \cap Dr_{i \in I(\beta)} F_i$ such that $(x^{-1}g)(J \cap I(\beta)) = y(J \cap I(\beta))$. Also $y(J_n) = 1$ so that $(xy)(J_n) = x(J_n) = g(J_n)$. Hence $\pi_J(xy) = \pi_J(g)$ and also $xy \in Dr\{F_i : i \in J_{n+1} \cup I(\beta)\} \leqslant Dr_{i \in I(\alpha)} F_i$. Thus condition (ii) is satisfied and $I(\alpha)$ is the required subset. □

Theorem 2.3. (Gorčakov [47]) If $G \in \mathcal{SDF}$, then G' is a direct product of countable groups.

Proof. We assume that $G \leqslant Dr_{i \in I} F_i$ and that the index set I is expressed as a union of sets $I(\alpha)$ as in Theorem 2.2. Write $G_\alpha = G \cap Dr_{i \in I(\alpha)} F_i$ so that $G = \bigcup_{\alpha<\rho} G_\alpha$ and $G' = \bigcup_{\alpha<\rho} G'_\alpha$. We also let $X_\alpha = G'_{\alpha+1} \cap Dr_{i \in J(\alpha)} F_i$; then each X_α is clearly countable and we shall prove that $G' \leqslant Dr_{\alpha<\rho} X_\alpha$. Since the sets $J(\alpha)$ are pairwise disjoint it is clear that the group generated by the X_α's is, in fact, their direct product. It is therefore sufficient for us to prove, by induction on α, that $G'_\alpha \leqslant Dr_{\alpha<\rho} X_\alpha$. Since $I(1) = \emptyset$, $G_1 = 1$ and so $G'_1 = 1 \leqslant Dr_{\alpha<\rho} X_\alpha$.

Suppose, therefore, that $G'_\beta \leqslant Dr_{\alpha<\rho} X_\alpha$, for each $\beta < \alpha$. If α is a limit ordinal, then $G_\alpha = \bigcup_{\beta<\alpha} G_\beta$ and so $G'_\alpha = \bigcup_{\beta<\alpha} G'_\beta \leqslant Dr_{\alpha<\rho} X_\alpha$. If $\alpha = \beta+1$ and x,y are elements of G_α, we write $x = x_1x_2$ and $y = y_1y_2$, where $x_1 = x(I(\beta))$, $x_2 = x(J(\beta))$, $y_1 = y(I(\beta))$ and $y_2 = y(J(\beta))$. Let J_1 be the finite subset $supp(x_1) \cup supp(y_1)$ of $I(\beta)$. By the definition of $I(\beta)$, there is an element $g \in G \cap Dr_{i \in I(\beta)} F_i$ such that $g(J_1) = x(J_1) = x_1(J_1)$ and hence $[g,y_1] = [x_1,y_1]$. Now let J_2 be the finite subset $supp(g) \cup supp(y_1)$ of $I(\beta)$; there is an element $h \in G \cap Dr_{i \in I(\beta)} F_i$ such that $h(J_2) = y(J_2) = y_1(J_2)$ and hence $[g,h] = [g,y_1] = [x_1,y_1]$. Therefore $[x,y] = [x_1,y_1][x_2,y_2] = [g,h][x_2,y_2]$ and so $[x_2,y_2] \in G'_\alpha$. But clearly $[x_2,y_2] \in Dr_{i \in J(\beta)} F_i$ and so $[x_2,y_2] \in X_\beta$. But, by induction $[g,h] \in G'_\beta \leqslant Dr_{\alpha<\rho} X_\alpha$ and so we have $[x,y] \in Dr_{\alpha<\rho} X_\alpha$. It follows that $G'_\alpha \leqslant Dr_{\alpha<\rho} X_\alpha$ and the proof is complete. □

Later in this chapter we shall be able to extend Theorem 2.3 to a more general result (Corollary 2.27). It is already possible to obtain a result about the central factor group in a wider class of groups; we shall see some indication of just how large this class is in this and the following chapter.

Theorem 2.4. (Gorčakov [47]) If $G \in \mathcal{QSDF}$, then G is isomorphic to a factor group H/K, where H is contained in a direct product $D = \mathrm{Dr}\, H_\alpha$ of countable periodic FC-groups such that if Z/K is the centre of H/K, then $Z = H \cap Z(D)$.

Hence $G/Z(G) \cong H/Z \cong HZ(D)/Z(D)$ is isomorphic to a subgroup of $\mathrm{Dr}(H_\alpha/Z(H_\alpha))$.

Proof. Let $G = U/V$, where U is contained in the direct product $\mathrm{Dr}_{i\in I}\, F_i$ of finite groups. Express I as a union of sets $I(\alpha), \alpha < \rho$, as given in Theorem 2.2. Let $U_\alpha = U \cap \mathrm{Dr}_{i\in J(\alpha)}\, F_i$, $V_\alpha = V \cap \mathrm{Dr}_{i\in J(\alpha)}\, F_i = V \cap U_\alpha$ and let $U_o = \mathrm{Dr}_{\alpha<\rho}\, U_\alpha \leqslant U$ and $V_o = \mathrm{Dr}_{\alpha<\rho}\, V_\alpha \leqslant V$.

Now let $P_\alpha = U(J(\alpha))$; then each P_α is countable and $U \leqslant \mathrm{Dr}_{\alpha<\rho}\, P_\alpha$. Thus $U/V_o \leqslant \mathrm{Dr}_{\alpha<\rho}\,(P_\alpha/V_\alpha)$. We write H_α for P_α/V_α, H for U/V_o and K for V/V_o; then $G \cong H/K$ with $H \leqslant \mathrm{Dr}_{\alpha<\rho}\, H_\alpha$.

Now let $Z/K = Z(H/K)$; then $Z = Y/V_o$ where $Y/V = Z(U/V)$. We show that $[Y,U] \leqslant V_o$. Let y and u be elements of Y and of $U \cap \mathrm{Dr}_{i\in I(\alpha)}\, F_i$. We prove by induction on α that $[y,u] \in V_o$ and we may clearly assume that α is a non-limit. So let $\alpha = \beta+1$ and write $y = y_1y_2$ and $u = u_1u_2$ where $y_1 = y(I(\beta))$, $y_2 = y(I-I(\beta))$, $u_1 = u(I(\beta))$ and $u_2 = u(J(\beta))$. If J is the finite subset $\mathrm{supp}(y_1) \cup \mathrm{supp}(u_1)$ of $I(\beta)$ then by the definition of $I(\beta)$ there is an element $x \in U \cap \mathrm{Dr}_{i\in I(\beta)}\, F_i$ such that $x(J) = u(J) = u_1(J)$ and hence $[y,x] = [y_1, u_1]$. Therefore $[y,u] = [y,x][y_2,u_2]$ and so $[y_2,u_2] \in [Y,U] \cap \mathrm{Dr}_{i\in J(\beta)} F_i \leqslant V \cap \mathrm{Dr}_{i\in J(\beta)}\, F_i = V_\beta$. But, by induction, $[y,x] \in V_o$ and hence $[y,u] = [y,x]\,[y_2,u_2] \in V_o$.

Since $[Y,U] \leqslant V_o$ we have $Y/V_o = Z(U/V_o)$; that is, $Z = Z(H)$ and so $Z(D) \cap H \leqslant Z$.

Also $[Y,P_\alpha]$ is the projection of $[Y,U]$ in $\mathrm{Dr}_{i\in J(\alpha)}\, F_i$ and, since $[Y,U] \leqslant V_o = \mathrm{Dr}_{\alpha<\rho}\, V_\alpha$, we have $[Y,P_\alpha] \leqslant V_\alpha$ and $[Y, \mathrm{Dr}_{\alpha<\rho}\, P_\alpha] \leqslant V_o$. Hence $[Z,\mathrm{Dr}_{\alpha<\rho}\, H_\alpha] = 1$ and so $Z \leqslant Z(D)$, as required. □

The first sufficient condition for a group to be an $\mathcal{SDF}$-group is Hall's Theorem on countable groups.

Theorem 2.5. (Hall [49]) If G is a countable residually finite periodic FC-group, then $G \in \mathcal{SDF}$.

Proof. The countable group G can be expressed as the union of an ascending chain of finite normal subgroups G_n,

$$1 = G_o \leq G_1 \leq \ldots \leq \bigcup_{n=1}^{\infty} G_n = G.$$

Since G is residually finite, there is a descending chain of normal subgroups of finite index in G,

$$G = N_o \geq N_1 \geq \ldots ,$$

such that $N_n \cap G_n = 1$. Hence $\bigcap_{n=1}^{\infty} N_n = 1$.

Now consider the normal subgroups $M_n = G_{n-1} N_n$ $(n \geq 1)$; clearly each M_n is a normal subgroup of finite index in G. We prove by induction on n that $\bigcap_{m \leq n} M_m \leq N_n$. Certainly $M_1 \leq N_1$ and so we may assume that $\bigcap_{m \leq n-1} M_m \leq N_{n-1}$. Then $\bigcap_{m \leq n} M_m \leq N_{n-1} \cap M_n = N_{n-1} \cap G_{n-1} N_n =$ $= (N_{n-1} \cap G_{n-1}) N_n = N_n$.

It follows that $\bigcap_{n=1}^{\infty} M_n = 1$ and there is therefore an embedding of G into $\prod_{n=1}^{\infty} (G/M_n)$ given by $g \mapsto (g M_1, g M_2, \ldots)$. Each element g of G is contained in some G_n and $G_n \leq M_m$, for all $m \geq n$, so that in this embedding the image of g is $(g M_1, \ldots, g M_n, 1, 1, \ldots) \in \mathrm{Dr}_{n=1}^{\infty} (G/M_n)$. □

In the correspondence between these results and results for abelian groups this corresponds to the well known theorem of Prüfer ([35], p.88) that every countable residually finite periodic abelian group is a direct product of cyclic groups. We shall see that to some extent it is the abelian part of an FC-group which prevents it from being embedded in a direct product of finite groups. It is therefore useful to recall some of the properties of residually finite abelian groups, in particular the concept of a basic subgroup. The details may be found in Chapters VI and XI of [35] but the main points for our purposes may be summarized as follows. Every residually finite abelian p-group G contains a *basic subgroup* $B = \mathrm{Dr}_{n=1}^{\infty} B_n$, where B_n is a direct product of cyclic groups of order p^n. The factor group G/B is a divisible group and G is isomorphic to a subgroup of the periodic subgroup of $\prod_{n=1}^{\infty} B_n$.

The following standard example of an abelian group which is not a direct product of cyclic groups shows that countability is necessary in Hall's

Theorem and is in some ways typical of uncountable FC-groups which are not in the class $\mathcal{SDF}$.

Example 2.6. Let G be the periodic subgroup of the cartesian product $\prod_{n=1}^{\infty} C_{p^n}$; then G is clearly residually finite and is uncountable. But $G \notin \mathcal{SDF}$, for if G were contained in a direct product $D = \mathrm{Dr}_{i\in I}^{\infty} F_i$ of finite groups, then its basic subgroup $B = \mathrm{Dr}_{n=1}^{\infty} C_{p^n}$ would be contained in a countable direct factor X of D. But G/B is divisible and so GX/X would be divisible contrary to D/X being residually finite.

In any FC-group G, the factor group G/Z(G) is a residually finite periodic FC-group and so we have the following Corollary to Hall's Theorem.

Corollary 2.7. If G is a countable FC-group, then $G/Z(G) \in \mathcal{SDF}$. □

Combining this with Theorem 2.4, we obtain

Corollary 2.8. (Gorčakov [46]) If $G \in \mathcal{QSDF}$, then $G/Z(G) \in \mathcal{SDF}$. □

Some sufficient conditions for G to be in the class $\mathcal{QSDF}$ will be seen later and, together with the above corollary, will produce some rather more striking results (e.g. Corollary 2.26).

It might be expected, since G/Z(G) is always residually finite and periodic, that G/Z(G) might always be in the class $\mathcal{SDF}$. The following example shows that this is not necessarily the case.

Example 2.9. (Gorčakov [42]) Let $A = \mathrm{Dr}_{n=1}^{\infty} \langle a_n : a_n^{p^n} = 1 \rangle$ and let B be the periodic subgroup of $\prod_{n=1}^{\infty} \langle b_n : b_n^{p^n} = 1 \rangle$. If C is the quasicyclic p-group $\langle c_1, c_2, \ldots : c_1^p = 1,\ c_{n+1}^p = c_n \rangle$, then we can form the split extension G of $A \times C$ by B such that C is central, $[a_n, b_n] = c_n$ and $[a_m, b_n] = 1$, for $m \neq n$. The group G is nilpotent of class two with $G' = Z(G) = C$. Since $G/C \cong A \times B$ and B is not a direct product of cyclic subgroups (Example 2.6) it is clear that $G/Z(G) \notin \mathcal{SDF}$. □

Corollary 2.8 shows that the group G constructed here is not in the class $\mathcal{QSDF}$. Simpler examples of groups not in the class $\mathcal{QSDF}$ will be given in the next chapter and we shall see a more direct argument showing that $G \notin \mathcal{QSDF}$.

It will be observed that the centre of G in this example is isomorphic to C_{p^∞} and so is not residually finite. In fact, we know of no example in which Z(G) does not contain a subgroup isomorphic to C_{p^∞}. Given information about Z(G) we can prove results about the abelian group $Z_2(G)/Z(G)$ by using Kulikov's criterion for an abelian group to be a direct product of cyclic groups (see [35], p.87).

Theorem 2.10. (Kulikov's Criterion) An abelian p-group A is isomorphic to a direct product of cyclic groups if and only if A is the union of an ascending chain of subgroups

$$1 = A_0 \leqslant A_1 \leqslant \ldots \leqslant \bigcup_{n=1}^{\infty} A_n = A ,$$

such that the non-trivial elements of A_n have height less than k_n. □

An ascending chain of this type will be called a *Kulikov chain.*

Theorem 2.11. (Tomkinson [104]) Let G be a periodic FC-group such that $[G, Z_2(G)]$ is a direct product of cyclic groups. Then, for each $n \geqslant 2$, $Z_n(G)/Z_{n-1}(G)$ is a direct product of cyclic groups.

Proof. Using induction on n, it is clearly sufficient to prove that Z_2/Z is a direct product of cyclic groups and to prove this we need only prove that each Sylow p-subgroup P/Z of Z_2/Z is a direct product of cyclic groups. Now [G,P] is a p-group contained in $[G, Z_2]$ and so has a Kulikov chain

$$1 = A_0 \leqslant A_1 \leqslant \ldots \leqslant \bigcup_{n=1}^{\infty} A_n = [G,P] ,$$

such that the non-trivial elements of A_n have height at most k_n in [G,P].

Let $B_n = C_P(G/A_n)$. Since G is an FC-group, each element $u \in P$ satisfies $[u,G] \leqslant A_n$, for some n, and so $P = \bigcup_{n=1}^{\infty} B_n$. Let $\bar{x}$ be a non-trivial element of B_n/Z and suppose that $\bar{x}$ has height h in P/Z. Writing $\bar{x} = xZ$, this means that there is an element $y \in P$ such that $y^{p^h} \equiv x \pmod{Z}$. Let $g \notin C_G(x)$ so that [x,g] is a non-trivial element of A_n. Then

$[y,g]^{p^h} = [y^{p^h},g] = [x,g]$ and so $h \leq k_n$. Thus the non-trivial elements of B_n/Z have height at most k_n in P/Z and we have a Kulikov chain

$$1 = B_o/Z \leq B_1/Z \leq \ldots \leq \bigcup_{n=1}^{\infty} (B_n/Z) = P/Z$$

of P/Z. Hence P/Z is a direct product of cyclic groups.□

If we weaken the condition on $Z(G)$ to just having no quasicyclic subgroup then we are unable to prove that Z_2/Z is a direct product of cyclic groups. However the following result does suggest that further progress in this direction may be possible.

Theorem 2.12. (Tomkinson [104]) Let G be an FC-group which contains no subgroup isomorphic to C_{p^∞}. Let P/Z be any p-subgroup of Z_2/Z and let B/Z be a basic subgroup of P/Z. Then $|B/Z| = |P/Z|$.

Proof. Suppose that, for each $x \in P$, there is a finite subgroup F_x/Z of B/Z, such that $C_G(x) \geq C_G(F_x)$. Then $x \in C_P(C_G(F_x))$ and, since $|G:C_G(F_x)|$ is finite, $C_P(C_G(F_x))/Z$ is also finite. Therefore P/Z is the union of the finite subgroups $C_P(C_G(F_x))/Z$. Since there are only $|B/Z|$ distinct subgroups F_x it follows that $|P/Z| = |B/Z|$.

We may therefore assume that P contains an element x_1 such that, for each finite subgroup F/Z of B/Z, there is an element $g_F \in G$ such that $[F,g_F] = 1$ but $[x_1,g_F] \neq 1$. We shall obtain a contradiction by constructing a subgroup isomorphic to C_{p^∞}.

There is a monomorphism θ: $P/Z \to \prod_{n=1}^{\infty} B_n$, where each B_n is a direct product of cyclic groups of order p^n, such that $\theta(B/Z) = \mathrm{Dr}_{n=1}^{\infty} B_n$. If $\bar{x}_1$ is the image of x_1 in P/Z then we may write

$$\theta(\bar{x}_1) = (b_1,b_2,\ldots) \in \prod_{n=1}^{\infty} B_n .$$

Since P/B is a divisible abelian p-group, there are elements $x_2,x_3,\ldots \in P$ such that $x_n^p x_{n-1}^{-1} \in B$, for each $n \geq 2$. These elements may be chosen so that

$$\theta(\bar{x}_1) = (d_1, d_2^p, \ldots, d_n^{p^{n-1}}, \ldots)$$

$$\theta(\bar{x}_n) = (1, \ldots, 1, d_n, d_{n+1}^p, \ldots),$$

where

$$d_1 = (b_1, \ldots, b_{r(1)}), d_n^{p^{n-1}} = (b_{r(n-1)+1}, \ldots, b_{r(n)}) \text{ and } d_n \in \mathrm{Dr}_{i=r(n-1)+1}^{r(n)} B_i .$$

Let $\bar{c}_n = \theta^{-1}(d_n)$ and let c_n be an element of B such that $c_n Z = \bar{c}_n$; then $x_1 \in c_1 c_2^p \ldots c_n^{p^{n-1}} P^{p^{r(n-1)+1}}$.

There is an element $y_1 \in G$ such that $[x_1, y_1] \neq 1$. Since $|P:C_P(y_1)|$ is finite, it follows that $[P^{p^k}, y_1] = 1$, for some k. Therefore there is an integer k(1) such that

$$[x_1, y_1] = [c_1 \ldots c_{k(1)}^{p^{k(1)-1}}, y_1] \neq 1.$$

By our choice of x_1, there is an element $y_2 \in C_G(\langle c_1, \ldots, c_{k(1)} \rangle) - C_G(x_1)$. As above there is an integer k(2) such that

$$[x_1, y_2] = [c_{k(1)+1}^{p^{k(1)}} \cdots c_{k(2)}^{p^{k(2)-1}}, y_2] \neq 1 .$$

Continuing in this way, we can find an infinite sequence of integers $k(1) < k(2) < \ldots$ and elements $y_1, y_2, \ldots$ such that

$$y_n \in C_G(\langle c_1, \ldots, c_{k(n-1)} \rangle), \text{ for all } n \geq 2,$$

and

$$[x_1, y_n] = [c_{k(n-1)+1}^{p^{k(n-1)}} \cdots c_{k(n)}^{p^{k(n)-1}}, y_n] \neq 1.$$

Since $[x_1, G]$ is finite, infinitely many of the $[x_1, y_n]$ must be equal to a_1, say. Therefore we have elements $g_{11}, g_{12}, \ldots$ and integers $s(1,1) \leq t(1,1) < s(1,2) \leq t(1,2) < \ldots$ such that

$$a_1 = [x_1, g_{11}] = [x_1, g_{12}] = \ldots ,$$

$$[x_1, g_{1n}] = [c_{s(1,n)}^{p^{s(1,n)-1}} \cdots c_{t(1,n)}^{p^{t(1,n)-1}}, g_{1n}],$$

$$g_{1n} \in C_G(<c_1, \ldots, c_{s(1,n)-1}>).$$

We use this as the first step in an induction argument. Assume that we have sets of elements

$$\{g_{11}, g_{12}, \ldots\} = S_1 \supseteq S_2 \supseteq \ldots \supseteq S_m = \{g_{m1}, g_{m2}, \ldots\}$$

and, for each $k = 1, \ldots, m$, we have integers $s(k,1) \leq t(k,1) < s(k,2) \leq t(k,2) < \ldots$ such that

$$a_k = [x_k, g_{k1}] = [x_k, g_{k2}] = \ldots,$$

$$[x_k, g_{kn}] = [c_{s(k,n)}^{p^{s(k,n)-k}} \cdots c_{t(k,n)}^{p^{t(k,n)-1}}, g_{kn}],$$

$$g_{kn} \in C_G(<c_1, \ldots, c_{s(k,n)-1}>).$$

Then $a_m^p = a_{m-1}, a_{m-1}^p = a_{m-2}, \ldots, a_2^p = a_1$. We show that a_{m+1} can be constructed by showing that there is an infinite subset $\{g_{m+1,1}, g_{m+1,2}, \ldots\} \subseteq S_m$ having the corresponding properties with k replaced by $m+1$.

Let $h_1 = g_{m1}$; then

$$[x_{m+1}, h_1] = [c_{s(m,1)}^{p^{s(m,1)-m-1}} \cdots c_{u(m,1)}^{p^{u(m,1)-m-1}}, h_1],$$

for some $u(m,1) \geq t(m,1)$.

Now let h_2 be the first g_{mn} for which $s(m,n) > u(m,1)$; then

$$[x_{m+1}, h_2] = [c_{s(m,n)}^{p^{s(m,n)-m-1}} \cdots c_{u(m,2)}^{p^{u(m,2)-m-1}}, h_2],$$

for some $u(m,2) \geq t(m,n)$. Then we can take h_3 to be the first g_{mn} for which $s(m,n) > u(m,2)$ and continuing in this way we obtain an infinite subset $\{h_1, h_2, \ldots\}$.

Since $h_i \in C_G(<c_1, \ldots, c_{s(m,n)-1}>)$ for some n, and $\bar{x}_{m+1}^p = \bar{c}_m^{-1}\bar{x}_m$, we have $[x_{m+1},h_i]^p = [x_m,h_i] = a_m$, for all but finitely many i. Since $[x_{m+1},G]$ is finite, infinitely many of the commutators $[x_{m+1},h_i]$ are equal to a_{m+1}, say, and $a_{m+1}^p = a_m$. Labelling this subset of $\{h_1,h_2,\ldots\}$ as $g_{m+1,1}, g_{m+2,1},\ldots$, we obtain the required subset S_{m+1}.

This shows that the construction of the elements a_m can be continued and so we can construct the subgroup $\langle a_1,a_2,\ldots\rangle \cong C_{p^\infty}$. □

The condition on Z_2/Z given in the above Theorem is a very strong condition and shows, for example, that Z_2/Z can not contain a subgroup isomorphic to that given in Example 2.6. These last two results do suggest that some stronger results may be possible for $G/Z(G)$ rather than just Z_2/Z and, in particular, positive answers to the following questions seem a possibility.

Question 2A. If G is an FC-group such that $G' \in \mathcal{SDF}$, is it necessarily true that $G/Z(G) \in \mathcal{SDF}$?

Question 2B. If G is an FC-group which contains no quasicyclic subgroup, is it necessarily true that $G/Z(G) \in \mathcal{SDF}$?

The next theorem we prove is the very important result of Gorčakov which corresponds to another result of Prüfer on abelian groups; namely that an abelian group of finite exponent is isomorphic to a direct product of cyclic groups ([35] , p.88).

Theorem 2.13. (Gorčakov [45]) *If the periodic FC-group G is a subgroup of the cartesian product $\Pi_{i\in I} F_i$ of isomorphic finite groups F_i, then $G \in \mathcal{SDF}$.*

The proof of this result is rather involved and we need first to introduce some notation. As before we denote the projection of G into F_i by $G(i)$. Since the group F_i has only finitely many subgroups, there are only finitely many possibilities, say $1 = X_{i,1}, X_{i,2},\ldots,X_{i,n} = F_i$ for the projections $G(i)$. For each $i = 1, \ldots,n$, define $I(r) = \{i\in I\colon G(i) = X_{i,r}\}$ then $G \leq \mathrm{Dr}_{r=1}^n G(I(r))$ and $G(I(r)) \leq \Pi_{i\in I(r)} X_{i,r}$. By induction on $|F_i|$ we may assume that $G(I(r)) \in \mathcal{SDF}$ for each $r < n$ and so we need only prove that

$G(I(n)) \in \mathcal{SDF}$. Thus we may assume that $G = G(I(n))$; that is, $I(n) = I$ and $G(i) = F_i$, for all $i \in I$. The centre of G is therefore $Z(G) = G \cap \Pi_{i\in I} Z(F_i)$ and so $G/Z(G)$ is isomorphic to a subgroup of $\Pi_{i\in I}(F_i/Z(F_i))$. These central factor groups will play an important part in the proof and we denote $G/Z(G)$ by $\bar{G}$ and $F_i/Z(F_i)$ by $\bar{F}_i$. Then $\bar{G} \lesssim \Pi_{i\in I} \bar{F}_i$ and for each subset J of I, we shall write $\bar{G}(J)$ for the projection of $\bar{G}$ in $\Pi_{i\in J} \bar{F}_i$ and if $\bar{g} \in \bar{G}$ then $\bar{g}(J)$ will denote the projection of the element $\bar{g}$ in $\Pi_{i\in J} \bar{F}_i$.

The subset J of I is called a *homogeneous* subset for G in I if $\bar{G}(J) \cong \bar{F}_i$. Any single element of I forms a homogeneous subset for G since $G(i) = F_i$ and hence $\bar{G}(i) = \bar{F}_i$. Clearly there are only $|\bar{F}_i|$ elements of G whose images in $\bar{G}(i)$ are all different. A homogeneous subset may be considered as a non-empty set J such that G still has only $|\bar{F}_i|$ elements with distinct images in the projection $\bar{G}(J)$. It is clear that any subset of a homogeneous subset is also homogeneous.

<u>Lemma 2.14.</u> <u>Every homogeneous subset is contained in a maximal homogeneous subset.</u>

<u>Proof.</u> By Zorn's Lemma, it is sufficient to show that the union of an ascending chain (J_λ) of homogeneous subsets is also homogeneous. Let $J = \bigcup J_\lambda$; if J is not homogeneous then there are $|\bar{F}_i| + 1$ elements $g_1,\dots,g_{n+1} \in G$ such that the projections $\bar{g}_1(J),\dots,\bar{g}_{n+1}(J)$ are distinct. It clearly follows that there is some λ such that $\bar{g}_1(J_\lambda),\dots,\bar{g}_{n+1}(J_\lambda)$ are distinct, contrary to J_λ being homogeneous. □

<u>Lemma 2.15.</u> <u>Distinct maximal homogeneous subsets are disjoint.</u>

<u>Proof.</u> Let J_1 and J_2 be homogeneous subsets with $J_1 \cap J_2 \neq \emptyset$ and let $j \in J_1 \cap J_2$. We can choose n elements $g_1,\dots,g_n \in G$ such that $\bar{F}_j = \{\bar{g}_1(j),\dots,\bar{g}_n(j)\}$. Since J_1 and J_2 are homogeneous, we therefore have $\bar{G}(J_1) = \{\bar{g}_1(J_1),\dots,\bar{g}_n(J_1)\}$ and $\bar{G}(J_2) = \{\bar{g}_1(J_2),\dots,\bar{g}_n(J_2)\}$. If g is any element of G, there is an i, $1 \leq i \leq n$, such that $\bar{g}(j) = \bar{g}_i(j)$. Hence $\bar{g}(J_1) = \bar{g}_i(J_1)$ and $\bar{g}(J_2) = \bar{g}_i(J_2)$ from which it follows that $\bar{g}(J_1 \cup J_2) = \bar{g}_i(J_1 \cup J_2)$. Thus $|\bar{G}(J_1 \cup J_2)| = |\bar{F}_i|$ and so $J_1 \cup J_2$ is homogeneous. □

These two lemmas show that the maximal homogeneous subsets for G form a partition of I. We refer to these maximal homogeneous subsets as the *homogeneous components* for G in I.

Lemma 2.16 If $J_\lambda, \lambda \in \Lambda$, are the homogeneous components for G in I and if I_1 is a subset of I, then the sets $J_\lambda \cap I_1, \lambda \in \Lambda$, are the homogeneous components for $G(I_1)$ in I_1.

Proof. Since $J_\lambda \cap I_1$ is a homogeneous subset for G it is clearly homogeneous for $G(I_1)$. If J is a homogeneous subset for $G(I_1)$ with $J_\lambda \cap I_1 \subseteq J \subseteq I_1$, then $|\overline{G(I_1)}(J)| = |\bar{F}_j|$. But $\overline{G(I_1)}(J) = \bar{G}(J)$ and so J is a homogeneous subset for G. Therefore $J = J_\lambda \cap I_1$ and $J_\lambda \cap I_1$ is a homogeneous component for $G(I_1)$. □

Lemma 2.17. Let J be a homogeneous component for G; then

(i) $G(J) = H \times T$, where H is finite and $T \leq Z(G(J))$,

(ii) there is a finite subset K of J such that $H \cong H(K)$.

Proof. (i) Since J is homogeneous, the centre of G(J) has finite index and so there is a finite subgroup H_1 of G(J) such that $G(J) = H_1 Z$, where Z is the centre of G(J). But Z has finite exponent and so, by Prüfer's Theorem, is a direct product of finite cyclic groups. Thus $H_1 \cap Z$ is contained in a finite direct factor H_2 of Z and $Z = H_2 \times T$, say. Letting $H = H_1 H_2$ we have $HT = H_1 Z = G(J)$ and $H \cap T = H \cap Z \cap T = H_2(H_1 \cap Z) \cap T = H_2 \cap T = 1$, so that $G(J) = H \times T$.

(ii) Letting π_i be the projection map from G into F_i, we have $G(J) = G/\bigcap_{i \in J} \operatorname{Ker} \pi_i$. Since H is a finite subgroup of G(J), there is a finite subset K of J such that $H \cap \bigcap_{i \in K} \operatorname{Ker} \pi_i$ is trivial and hence $H \cong H(K)$. □

The proof of Gorčakov's Theorem is basically a proof by induction on the order of F_i. We therefore need to consider subsets of I for which the corresponding projection of G is embeddable in a cartesian product of isomorphic finite groups of order less than $|F_i|$. We define a subset J of I to be a *reducing subset* for G if G(J) is embeddable in a cartesian product of finite groups each having order less than $|F_i|$. (We do not insist that the finite groups in this product are isomorphic). If $J = \bigcup J_\lambda$ and each J_λ is a reducing subset for G, then since G(J) is isomorphic to a subgroup of $\prod G(J_\lambda)$ it is clear that J is also a reducing subset. Thus the index set I contains a unique maximal reducing subset for G which we call *the reducing component* for G in I.

Since there are only finitely many non-isomorphic groups of order less than $|F_i|$ it follows that G(J) can be embedded in a finite direct product of groups G_n, each of which is embeddable in a cartesian product of isomorphic finite groups of order less than $|F_i|$. Then an induction argument shows that if J is the reducing component for G in I, then $G(J) \in \mathcal{SDF}$.

<u>Proof of Gorčakov's Theorem</u>. We use induction on $|F_i|$. By Lemmas 2.14 and 2.15, we can write I as the disjoint union of homogeneous components $J_\lambda, \lambda \in \Lambda$. By Lemma 2.17, $G(J_\lambda) = H_\lambda \times T_\lambda$, where T_λ is central, H_λ is finite and there is a finite subset K_λ of J_λ such that $H_\lambda(K_\lambda) \cong H_\lambda$. For each $\lambda \in \Lambda$, let $M_\lambda = \bigcap_{i \in K_\lambda} \operatorname{Ker} \pi_i$ so that $G/M_\lambda \cong G(K_\lambda)$ and let N_λ be the complete inverse image of H_λ in G. Then $M_\lambda \cap N_\lambda = \bigcap_{i \in J_\lambda} \operatorname{Ker} \pi_i$ (that is, $G/(M_\lambda \cap N_\lambda) \cong G(J_\lambda)$) and so $\bigcap_{\lambda \in \Lambda} M_\lambda \cap \bigcap_{\lambda \in \Lambda} N_\lambda = 1$. Therefore G can be embedded in the direct product $(G/\bigcap_{\lambda \in \Lambda} M_\lambda) \times (G/\bigcap_{\lambda \in \Lambda} N_\lambda)$.

But $G/N_\lambda \cong T_\lambda$ and so $G/\bigcap_{\lambda \in \Lambda} N_\lambda$ is an abelian group of finite exponent and so is in the class $\mathcal{SDF}$.

Therefore, we need only consider $G/\bigcap_{\lambda \in \Lambda} M_\lambda = G(K)$, where $K = \bigcup_{\lambda \in \Lambda} K_\lambda$. By Lemma 2.16, the finite sets $K_\lambda, \lambda \in \Lambda$, are the homogeneous components for G(K) in K. Let K_1 be the reducing component for G(K) in K and let $K_2 = K - K_1$. Then G(K) can be embedded in the direct product $G(K_1) \times G(K_2)$. By the remarks about the reducing component, the induction hypothesis shows that $G(K_1) \in \mathcal{SDF}$ and it only remains to prove that $G(K_2) \in \mathcal{SDF}$. By Lemma 2.16 again, the homogeneous components for $G(K_2)$ in K_2 are the finite sets $K_2 \cap K_\lambda, \lambda \in \Lambda$, and also there is no reducing subset for $G(K_2)$ in K_2. We have therefore reduced our considerations to the following situation:

<u>The periodic FC-group G is a subgroup of the cartesian product</u> $\prod_{i \in I} F_i$ <u>of isomorphic finite groups</u> F_i. <u>The homogeneous components for G in I are all finite and there is no reducing subset for G in I.</u> (*)

Before proving the main lemma which will complete the proof of Gorčakov's Theorem, we make one further definition. Unlike the other definitions introduced in this proof, this will be of great importance in later results.

Let $G_i, i \in I$, be periodic groups with centres Z_i. We define the *centrally restricted product* of the groups G_i to be the subgroup of $\Pi_{i\in I} G_i$ consisting of all elements of finite order with only finitely many components not belonging to Z_i. We denote this product by $Zr_{i\in I} G_i$. This group can also be described as the product of the periodic subgroup of $\Pi_{i\in I} Z_i$ and the direct product $Dr_{i\in I} G_i$.

Lemma 2.18. Let $G \leq Zr_{i\in I} F_i$, where the F_i are isomorphic finite groups. Then $G \in S\mathcal{DF}$.

Proof. We can write $Zr_{i\in I} F_i = Z D$, where $Z = \Pi_{i\in I} Z(F_i)$ and $D = Dr_{i\in I} F_i$. But Z has finite exponent and $Z \cap D$ is a pure subgroup of Z so that $Z \cap D$ is a direct factor of Z ([35], p.117). Writing $Z = (Z \cap D) \times E$, we obtain $ZD = D \times E$, $D \in \mathcal{DF}$ and E, being abelian of finite exponent, is also in the class $\mathcal{DF}$. Thus $ZD \in \mathcal{DF}$ and so $G \in S\mathcal{DF}$. □

To complete the proof of Gorčakov's Theorem it is now sufficient for us to prove that the group G described in (*) is contained in the centrally restricted product $Zr_{i\in I} F_i$. This is given in the following result.

Lemma 2.19. Let the periodic FC-group G be a subgroup of the cartesian product $\Pi_{i\in I} F_i$ of isomorphic finite groups F_i. Suppose also that all the homogeneous components for G in I are finite. If some element of G has infinitely many non-central components, then I contains an infinite reducing subset for G.

Proof. (A) If H is a finite normal subgroup of G, then $H(i)$ is a proper subgroup of F_i, for all but finitely many $i \in I$.

Let $J = \{i \in I: H(i) = F_i\}$; we must show that J is finite. Let $H_1 = H(J)$ and $G_1 = G(J)$; then H_1 is a finite normal subgroup of G_1 and $H_1(i) = H(i) = F_i$, for each $i \in J$.

Now $C_{G_1}(H_1) = G_1 \cap \Pi_{i\in J} C_{F_i}(H_1(i)) = G_1 \cap \Pi_{i\in J} Z_i = Z(G_1)$ and so $G_1/Z(G_1)$ is finite. Thus $\bar{G}_1 = G_1/Z(G_1)$ has only finitely many subgroups $X_1, \ldots, X_r$, say. Let $\bar{\pi}_i$ be the projection map of $\bar{G}_1$ into $\bar{F}_i$ and, for each $s = 1, \ldots, r$, let $J_s = \{i \in J: \operatorname{Ker} \bar{\pi}_i = X_s\}$. Then, for each $i \in J_s$, $\bar{G}_1(i) \cong \bar{G}_1/X_s \cong \bar{G}_1(J_s)$. Thus J_s is a homogeneous subset for G_1 and hence also for G. Therefore each J_s is finite and so $J = J_1 \cup \ldots \cup J_r$ is also finite, as required.

(B) <u>Suppose that I contains an infinite subset I_1 with the property:</u>

<u>$\bar{G}(I_1) \leq (\mathrm{Dr}_{i\in I_1} \bar{F}_i)\bar{H}$, where $\bar{H}$ is a finite normal subgroup of $\bar{G}(I_1)$ such that $\bar{H}(i) \neq 1$, for all $i \in I_1$. Then I_1 contains an infinite reducing subset for G.</u> (**)

Since the homogeneous components for $G(I_1)$ in I_1 are finite by Lemma 2.16, we may clearly assume that $I_1 = I$. Also the condition (**) only applies to the factor group $\bar{G} = G/Z(G)$ and so we may replace G by $G(\prod_{i\in I} Z_i)$ and so assume that $G \geq \prod_{i\in I} Z_i = Z$ and hence that $Z(G) = Z$. Writing $\bar{D} = \mathrm{Dr}_{i\in I} \bar{F}_i$, we now have $\bar{G} \leq \bar{D}\bar{H}$ and so $\bar{G} = (\bar{G} \cap \bar{D})\bar{H}$. If E and H are the inverse images of $\bar{G} \cap \bar{D}$ and $\bar{H}$, then we have $G = EH$.

Let $D = G \cap (\mathrm{Dr}_{i\in I} F_i) = E \cap (\mathrm{Dr}_{i\in I} F_i)$; then $E = DZ$. Also H/Z is finite and so G has a finite normal subgroup F such that $H = FZ$. The subgroup F can be chosen to contain the diagonal subgroup of $\prod_{i\in I} Z_i$ and so we may assume that $F(i) \geq Z_i$. Note also that $\bar{F}(i) = \bar{H}(i) \neq 1$ and so $F(i) > Z_i$, for each $i \in I$. We now have $G = EH = DFZ$.

By (A), the set $K_1 = \{i \in I: F(i) = F_i\}$ is finite. Also $K_2 = \operatorname{supp}(D \cap F)$ is finite. Let $J = I - (K_1 \cup K_2)$; then $G(J) = D(J)\, F(J)\, Z(J)$ and, since $K_1 \cup K_2$ is finite, $D(J) = G(J) \cap \mathrm{Dr}_{i\in J} F_i$. Also $F(J)$ is a finite normal subgroup of $G(J)$ and $D(J) \cap F(J) = 1$. Thus $G(J) = (D(J) \times F(J))\, Z(J)$.

To simplify the notation we shall now assume that the index set I is replaced by the set J. We therefore have $G = (D \times F)Z$, where D and Z are as defined previously and F is a finite normal subgroup of G such that $Z_i < F(i) < F_i$, for each $i \in I$. Also, letting $X = Z(F)$ and $Y = Z(D)$, we have $(D \times F) \cap Z = Y \times X$.

Writing A_p for the Sylow p-subgroup of an abelian group A, we have $Z_p = \prod_{i\in I} (Z_i)_p$ and $Y_p = \mathrm{Dr}_{i\in I}(Z_i)_p$ so that Y_p is a pure subgroup of Z_p. The finite subgroup $(Y_p \times X_p)/Y_p$ of the abelian group Z_p/Y_p of finite exponent is contained in a finite direct factor T_p/Y_p. This will be a pure subgroup of Z_p/Y_p and hence T_p is a pure subgroup of Z_p ([35] ,p.115). Therefore T_p is a direct factor of Z_p and we can write $Z_p = T_p \times U_p$. But also Y_p, being a pure subgroup of Z_p, is also pure in T_p and so is a direct factor of T_p. Therefore there is a finite subgroup V_p such that $T_p = Y_p \times V_p$. Let $W_p = X_p V_p \leq T_p$. Then W_p is finite and $T_p = Y_p W_p$ so that $Z_p = Y_p W_p \times U_p$.

Now let $W = F(\mathrm{Dr}_p W_p)$ and $U = \mathrm{Dr}_p U_p$. Then W is finite and $G = DFZ = DWU$.

Also $[D,W] \leq [D,FZ] = [D,F] = 1$ and so $Z(DW) = Z(D)\, Z(W) = Y\, X\, (\mathrm{Dr}_p W_p)$ $= \mathrm{Dr}_p Y_p W_p$. Thus $DW \cap U = Z(DW) \cap U = (\mathrm{Dr}_p Y_p W_p) \cap U = 1$, since $Y_p W_p \cap U_p = 1$. We therefore have $G = DW \times U$.

Now $D \cap W$ is a finite subgroup of $D = G \cap \mathrm{Dr}_{i \in I} F_i$ and so supp $(D \cap W)$ is finite. Let $J = I$ - supp $(D \cap W)$; then

$$G(J) = D(J) \times W(J) \times U(J).$$

For, let $dwu \in G$ be such that $(dwu)(J) = 1$. Then, since $I - J$ is finite, $dwu \in G \cap \mathrm{Dr}_{i \in I} F_i = D$. Thus $u \in DW \cap U = 1$ and so $dw \in D$. Therefore $w \in D \cap W$ and hence $w(J) = 1$ and consequently $d(J) = 1$.

For each $i \in J$, $\bar{W}(i) = \bar{F}(i) < \bar{F}_i$ (since $Z_i \leq F(i) < F_i$) and so $W(i) < F_i$. Also $\bar{W}(i) \neq 1$, since $Z_i < F(i)$, and so since $[D(i), W(i)] = 1$, we have $D(i) < F_i$. Since $U \leq Z$ it is clear that $U(i) \leq Z_i < F_i$ and hence J is the required reducing subset for G.

(C) <u>Completion of the proof</u>. We show, by induction on $|F_i|$, that there is a set I_1 satisfying the conditions of (B).

There is an element $x \in G$ such that $x(i) \notin Z_i$ for infinitely many $i \in I$. Replacing I by the appropriate subset we may assume that $x(i) \notin Z_i$, for all $i \in I$.

We say that two elements g and h of G are *parallel on a subset* J of I if $g(i) \neq h(i)$, for all $i \in J$. If S is a set of elements of G which are pairwise parallel on some subset of I, then clearly $|S| \leqslant |F_i|$ and so we can choose a set $\{x_1, \ldots, x_t\}$ maximal with respect to $x_1, \ldots, x_t$ being pairwise parallel on some infinite subset of I. We may clearly replace I by that subset so that $x_1, \ldots, x_t$ are pairwise parallel on I but no set of $t + 1$ elements is pairwise parallel on any infinite subset of I. Note also that by multiplying by a suitable element of G, we may assume that $x \notin \{x_1, \ldots, x_t\}$.

The set $J = \{i \in I: x(i) \neq x_j(i)$, for all $j = 1,\dots,t\}$ is then finite. Replacing I by $I - J$ we may now assume that, for each $i \in I$, $x(i) = x_j(i)$, for some $j = j(i) = 1,\dots,t$.

Let H be the finite normal subgroup $\langle x, x_1, \dots, x_t\rangle^G$. Since H has only finitely many subgroups, there will be an infinite subset K of I such that the $H \cap \operatorname{Ker} \pi_i$ are the same for all $i \in K$. We now replace I by K so that if $h \in H$ and $h(i) = 1$ for some $i \in I$, then $h(i) = 1$ for all $i \in I$ and so $h = 1$.

Let $g \in G$; then $g(i) = x_j(i)$, for some $i \in I$, $j \in \{1,\dots,t\}$. So $(g x_j^{-1})(i) = 1$ and, for any $h \in H$, $[h, g x_j^{-1}](i) = 1$. But $[h, g x_j^{-1}] \in H$ and so $[h, g x_j^{-1}] = 1$. Therefore $g x_j^{-1} \in C_G(H)$ and $g \in H\, C_G(H)$; hence $G = H\, C_G(H)$.

Since $x \in H$ and $x(i) \notin Z_i$, we have $C_G(H)(i) < F_i$, for all $i \in I$.

If $C_G(H) \lesssim \mathrm{Zr}_{i\in I}\, F_i$, then $\bar{G} = \bar{H}\, \overline{C_G(H)} \lesssim \bar{H}(\mathrm{Dr}_{i\in I}\, \bar{F}_i)$, the condition given in (B). If $C_G(H) \not\lesssim \mathrm{Zr}_{i\in I}\, F_i$, then there is an element $y \in C_G(H)$ with infinitely many non-central components. We can choose an infinite subset I_2 of I such that $y(i) \notin Z_i$, for all $i \in I_2$, and such that the projections $C_G(H)(i) = E_i$ are all isomorphic for $i \in I_2$. Since $|C_G(H)(i)| = |E_i| < |F_i|$, we can use the induction hypothesis to deduce that I_2 has an infinite subset I_1 such that $\overline{C_G(H)(I_1)} \lesssim (\mathrm{Dr}_{i\in I_1}\, \bar{E}_i)\, \bar{H}_1$, where $\bar{H}_1$ is a finite normal subgroup of $\overline{C_G(H)(I_1)}$ such that $\bar{H}_1(i) \neq 1$, for all $i \in I_1$. It follows that

$$\bar{G}(I_1) = \overline{C_G(H)(I_1)}\bar{H}(I_1) \lesssim (\mathrm{Dr}_{i\in I_1}\, \bar{E}_i)\, \bar{H}_1\, \bar{H}(I_1) \lesssim (\mathrm{Dr}_{i\in I_1}\, \bar{F}_i)\, \bar{H}_1\, \bar{H}(I_1),$$

and so I_1 satisfies the conditions of (B).

This completes the proof of Lemma 2.19 and hence the proof of Gorčakov's Theorem 2.13. □

Our main aim now is to make use of Gorčakov's Theorem to characterize completely the residually finite periodic FC-groups as subgroups of centrally restricted products of finite groups. We shall then be able to deduce a number of sufficient conditions for a group to be embeddable in a direct product of finite groups.

The first lemma, which is an extension of the method used by Hall in Theorem 2.5, enables us to consider a different set of normal subgroups intersecting in the identity.

<u>Lemma 2.20.</u> (Gorčakov [42]) <u>Let ρ be a limit ordinal and let $(M_\alpha : \alpha < \rho)$ be a descending chain of normal subgroups of the group G. If H_α, $\alpha < \rho$, are subgroups of G such that $M_\alpha \cap H_\alpha = 1$, then</u>

$$M_1 \cap (\bigcap_{\alpha<\rho} H_\alpha M_{\alpha+1}) = \bigcap_{\alpha<\rho} M_\alpha .$$

Proof. The inclusion in one direction is clear and so we prove, by induction on β, that $M_1 \cap (\bigcap_{\alpha<\beta} H_\alpha M_{\alpha+1}) \leq \bigcap_{\alpha<\beta} M_{\alpha+1}$. We assume the result for all ordinals $\gamma < \beta$. If β is a limit ordinal, then $M_1 \cap (\bigcap_{\alpha<\beta} H_\alpha M_{\alpha+1}) \leq$ $\leq \bigcap_{\gamma<\beta}(\bigcap_{\alpha<\gamma} M_{\alpha+1}) = \bigcap_{\alpha<\beta} M_{\alpha+1}$. If β is a non-limit, then $\beta = \gamma+1$ and $M_1 \cap (\bigcap_{\alpha<\beta} H_\alpha M_{\alpha+1}) \leq \bigcap_{\alpha<\gamma} M_{\alpha+1} \cap H_\gamma M_{\gamma+1} \leq M_\gamma \cap H_\gamma M_{\gamma+1} = (M_\gamma \cap H_\gamma) M_{\gamma+1} = M_\beta =$ $= \bigcap_{\alpha<\beta} M_{\alpha+1}$. □

Lemma 2.21. (Gorčakov [42] , Tomkinson [104]) Let $N_\alpha, \alpha < \rho$, be normal subgroups of the periodic FC-group G such that $\bigcap_{\alpha<\rho} N_\alpha = 1$ and let $C_\alpha = C_G(G/N_\alpha)$. If $(H_\alpha : \alpha < \rho)$ is an ascending chain of normal subgroups of G such that $[H_\alpha, G] \leq N_\beta$, for all $\beta \geq \alpha$, and $H_{\alpha+2} C_\alpha = G$, for all $\alpha < \rho$, then $G \leq Zr_{\alpha<\rho} (G/N_\alpha)$.

Proof. Let $x \in G$ and suppose that, in the embedding $G \to \Pi_{\alpha<\rho}(G/N_\alpha)$, x has infinitely many non-central components. Then there are infinitely many ordinals

$$\alpha(1) < \alpha(2) < \ldots$$

such that $x \notin C_{\alpha(i)}$. Then there is an element $h_i \in H_{\alpha(i)+2}$ such that $[h_i, x] \notin N_{\alpha(i)}$. But $[h_i, x] \in [H_{\alpha(i)+2}, G] \leq N_\beta$, for all $\beta \geq \alpha(i)+2$. In particular $[h_{2k}, x] \in N_{\alpha(2k+2)}$, since $\alpha(2k+2) \geq \alpha(2k)+2$, but $[h_{2k}, x] \notin N_{\alpha(2k)}$. Thus the commutators $[h_{2k}, x]$, $k = 1,2,\ldots$, are all distinct contrary to G being an FC-group. □

Lemma 2.22. (Gorčakov [44], Hartley [51]) If the periodic FC-group G is a subgroup of the cartesian product $\Pi_{i\in I} F_i$ of finite groups F_i and I is infinite, then $|G/Z(G)| \leq |I|$.

Proof. Let Λ be the set of finite subsets of I so that $|\Lambda| = |I|$ and, for each $\lambda \in \Lambda$, let $N_\lambda = G \cap \Pi_{i\notin\lambda} F_i$ so that G/N_λ is finite and $\bigcap_{\lambda\in\Lambda} N_\lambda = 1$. There is a finite subgroup X such that $G = N_\lambda X$ and so $Z(G) = C_G(N_\lambda) \cap C_G(X)$. Since $|G:C_G(X)|$ is finite, we have $C_G(N_\lambda)/Z(G)$ is finite, for each $\lambda \in \Lambda$.

Let $g \in G$; then g is contained in some finite normal subgroup F of G.

There is a $\lambda \in \Lambda$ such that $N_\lambda \cap F = 1$ and so $g \in F \leq C_G(N_\lambda)$. Thus $G/Z(G)$ is generated by the finite normal subgroups $C_G(N_\lambda)/Z(G)$ and the result follows.□

Lemma 2.23. *If* $G \leq Zr_{i\in I}F_i$ *and, for each* $i \in I$, $F_i \leq Zr_{j\in J(i)}E_j$ *and the projection of* F_i *on* E_j *is the whole of* E_j, *then* $G \leq Zr_{j\in J}E_j$, *where* $J = \bigcup_{i\in I} J(i)$.

Proof. We have $G \leq Zr_{i\in I}F_i \leq \Pi_{j\in J} E_j$. If $g \in G$, then only finitely many components $g(i)$ of g in ΠF_i are non-central. Each of these has only finitely many non-central components in ΠE_j. If $g(i)$ is central, then the projection of $g(i)$ in E_j is central, since the projection of $G(i)$ in E_j is the whole of E_j. Hence every component of $g(i)$ is central and so g has only finitely many non-central components in ΠE_j. □

We are now able to obtain the characterization of residually finite periodic FC-groups which will lead to a number of sufficient conditions for a group to be embedded in a direct product of finite groups.

Theorem 2.24. (Tomkinson [106]) *A group* G *is a residually finite periodic FC-group if and only if* G *is isomorphic to a subgroup of a centrally restricted product of finite groups.*

Proof. It is clear that a centrally restricted product of finite groups is a residually finite periodic FC-group and so we have only to consider the converse in which we assume that G is a residually finite periodic FC-group.

Let $X_1, X_2, \ldots$ be an enumeration of a set of representatives for the isomorphism types of all finite groups. Let

$$L_n = \bigcap \{N : N \triangleleft G,\ G/N \cong X_n\} ;$$

then $\bigcap_{n=1}^{\infty} L_n = 1$ and so $G \leq \Pi_{n=1}^{\infty} (G/L_n)$. But G/L_n is isomorphic to a subgroup of a cartesian product of groups each isomorphic to X_n and so, by Gorčakov's Theorem, G/L_n is isomorphic to a subgroup of $Dr_{i\in I(n)} F_i$, where the F_i are finite groups. We may therefore embed G in $\Pi_{i\in I} F_i$, where $I = \bigcup_{n=1}^{\infty} I(n)$, in such a way that each element of G has countable support.

We now prove the theorem by induction on the cardinality of this index set I.

(A) I countable. Writing $G \leq \prod_{n=1}^{\infty} F_n$, let $G_n = G \cap \prod_{m>n} F_m$ so that G/G_n is finite and

$$G \geq G_1 \geq \ldots \geq \bigcap_{n=1}^{\infty} G_n = 1.$$

We define subgroups H_n, M_n of G inductively as follows. Let $H_1 = 1$, $M_1 = G_1$ and for $n > 1$ define H_n to be a finite normal subgroup of G containing H_{n-1} such that $H_n M_{n-1} = G$; define M_n to be the first of the G_m such that $G_m \lneq M_{n-1}$ and $G_m \cap H_n = 1$.

Let $N_1 = M_1$ and $N_n = M_{n+1} H_n$, for $n > 1$. By Lemma 2.20, $\bigcap_{n=1}^{\infty} N_n = 1$ and by Lemma 2.21, $G \leq Zr_{n=1}^{\infty} (G/N_n)$

(B) I uncountable. We may assume that I consists of all ordinals $\alpha<\rho$, where ρ is the least ordinal of cardinality $|I|$. We partition I into disjoint sets J_α, $\alpha<\rho$, such that

(i) $\alpha \in \bigcup_{\beta\leq\alpha} J_\beta$

(ii) $|\bigcup_{\beta\leq\alpha} J_\beta| \leq \max(\aleph_0, |\alpha|)$,

(iii) if $H_\alpha = G \cap \Pi\{G_\gamma : \gamma \in \bigcup_{\beta<\alpha} J_\beta\}$, $M_\alpha = G \cap \Pi\{G_\gamma : \gamma \notin \bigcup_{\beta<\alpha} J_\beta\}$

and $Z_\alpha = C_G(G/M_\alpha)$, then $H_{\alpha+1} Z_\alpha = G$.

Suppose that we have defined the sets J_β for all $\beta < \alpha$, satisfying the conditions

(i)′ $\delta \in \bigcup_{\gamma\leq\delta} A_\gamma$, for all $\delta < \alpha$

(ii)′ $|\bigcup_{\gamma\leq\delta} A_\gamma| \leq \max(\aleph_0, |\delta|)$, for all $\delta < \alpha$,

(iii)′ $H_{\delta+1} Z_\delta = G$, for all δ such that $\delta+1 < \alpha$.

If α is a limit ordinal, let

$$J_\alpha = \begin{cases} \{\alpha\}, \text{ if } \alpha \notin \bigcup_{\beta<\alpha} J_\beta , \\ \emptyset \quad , \text{ if } \alpha \in \bigcup_{\beta<\alpha} J_\beta . \end{cases}$$

Then clearly conditions (i)′, (ii)′ hold for $\delta = \alpha$ and (iii)′ holds for all $\delta < \alpha$.

If α is a non-limit, let $\alpha = \varepsilon + 1$. The factor group G/M_ε is isomorphic to a subgroup of $\Pi\{F_\gamma : \gamma \in \bigcup_{\beta<\varepsilon} J_\beta\}$ and so, by Lemma 2.22, $|G/Z_\varepsilon| \leq$

$\max(\aleph_0, |\varepsilon|)$.

There is a subgroup S_ε of G such that $S_\varepsilon Z_\varepsilon = G$ and $|S_\varepsilon| \leq \max(\aleph_0, |\varepsilon|)$. Let $K_\varepsilon = \{\gamma : S_\varepsilon$ has non-trivial projection in $F_\gamma\}$ so that $|K_\varepsilon| \leq \max(\aleph_0, |\varepsilon|)$ and define

$$J_\alpha = (\{\alpha\} \cup K_\varepsilon) - (\bigcup_{\beta<\alpha} J_\beta).$$

Then $|\bigcup_{\beta\leq\alpha} J_\beta| \leq \max(\aleph_0, |\alpha|)$ and $H_\alpha Z_\varepsilon = G$. Thus (i)′ and (ii)′ hold for $\delta = \alpha$ and (iii)′ holds for all $\delta < \alpha$.

This completes the partitioning of the index set I.

Now define

$$N_\alpha = G \cap \Pi\{F_\beta : \beta \notin J_\alpha\} \geq M_{\alpha+1}.$$

If $C_\alpha = C_G(G/N_\alpha)$, then $H_{\alpha+2} C_\alpha \geq H_{\alpha+2} Z_{\alpha+1} = G$. Thus Lemma 2.21 shows that $G \leq Zr_{\alpha<\rho}(G/N_\alpha)$.

The group G/N_α is isomorphic to a subgroup of $\Pi\{F_\beta : \beta \in J_\alpha\}$ in such a way that each element has finite support. Also $|J_\alpha| < |I|$ and so, by induction, each G/N_α can be embedded in a centrally restricted product of finite groups. By Lemma 2.23, G itself is isomorphic to a subgroup of a centrally restricted product of finite groups. □

We can now deduce a series of conditions for a group to be embeddable in a direct product of finite groups. Most of these results were known before the theorem proved above.

Throughout these results we shall assume that a residually finite periodic FC-group G is contained in a centrally restricted product $Zr_{i\in I} F_i$ of finite groups. We write $D = Dr_{i\in I} F_i$ and $Z = \Pi_{i\in I} Z(F_i) = Z(Zr_{i\in I} F_i)$ so that $G \leq ZD$.

<u>Corollary 2.25.</u> (Gorčakov [44]) <u>A residually finite periodic FC-group G is in the class</u> $\mathcal{QSDF}$.

<u>Proof.</u> Using the notation above, ZD is isomorphic to the direct product of Z and D with the subgroup $Z \cap D$ amalgamated. The periodic abelian group Z is isomorphic to a subgroup of a direct product of quasicyclic groups and

a quasicyclic group C_{p^∞} is isomorphic to a factor group of the direct product $\mathrm{Dr}_{n=1}^{\infty} C_{p^n}$. Hence $Z \in \mathcal{SD}(\mathcal{QDF}) \subseteq \mathcal{QSDF}$. Also $D \in \mathcal{DF}$ and so $Z \times D \in \mathcal{QSDF}$ and $ZD \in \mathcal{QSDF}$. Hence $G \in \mathcal{QSDF}$. □

If G is any FC-group, then $G/Z_n(G)$ is a residually finite periodic FC-group, for each $n \geqslant 1$. Hence $G/Z_n(G) \in \mathcal{QSDF}$ and applying Corollary 2.8, we obtain

Corollary 2.26. (Gorčakov [46]) If G is any FC-group, then $G/Z_n(G) \in \mathcal{SDF}$, for all $n \geqslant 2$. If G is a residually finite periodic FC-group, then $G/Z(G) \in \mathcal{SDF}$. □

Corollary 2.27. (Tomkinson [104] and [106]) If G is a residually finite periodic FC-group then G' is a direct product of countable groups and hence $G' \in \mathcal{SDF}$.

Proof. Since $G' = (GZ)'$ and $GZ = (GZ \cap D)Z$, we have $G' = (GZ \cap D)'$. But $GZ \cap D \in \mathcal{SDF}$ and so the result follows from Theorem 2.3. □

Again noting that $G/Z(G)$ is always residually finite and periodic we have the following striking result.

Corollary 2.28. (Tomkinson [106]) If G is any FC-group, then $G'Z(G)/Z(G)$ is a direct product of countable $\mathcal{SDF}$-groups. □

Corollary 2.29. (Tomkinson [106]) Let G be a residually finite periodic FC-group with G/G' having finite exponent. Then $G \in \mathcal{SDF}$.

Proof. We consider the group $GD = (Z \cap GD)D$. The subgroup $Z \cap D$ is a pure subgroup of $Z \cap GD$ and $(Z \cap GD)/(Z \cap D) \cong GD/D \cong G/(G \cap D)$ which is abelian and hence has finite exponent. Thus $Z \cap D$ is a direct factor of $Z \cap GD$ which may be written as $E \times (Z \cap D)$ ([35], p.120). Therefore $GD = (Z \cap GD)D = ED$ and $E \cap D = E \cap (Z \cap D) = 1$. Thus $GD = E \times D \in \mathcal{DF}$ and hence $G \in \mathcal{SDF}$. □

It will have been noticed in the proof of Theorem 2.24 that Gorčakov's

Theorem was only used to prove that a residually finite periodic FC-group G can be embedded in $\Pi_{i\in I}F_i$ in such a way that each element of G has countable support. This is apparently a much weaker result than Gorčakov's Theorem and it is possible that a simpler proof could be found. This would be useful as Gorčakov's Theorem could then be deduced from Theorem 2.24 (for example, Corollary 2.29). We record this as our next open problem.

<u>Question 2C</u>. Is there a direct (simpler) proof that a residually finite periodic FC-group G can be embedded in a cartesian product $\Pi_{i\in I}F_i$ of finite groups such that each element of G has countable support?

A number of the results given in this Chapter say that a residually finite FC-group G will be in the class $\mathcal{SDF}$ if certain abelian sections are well-behaved (for example, Theorems 2.5 and 2.13 and Corollary 2.29) or if certain abelian sections are removed (for example, Corollaries 2.8 and 2.27). This way of looking at these results suggests a number of questions of which we mention two.

<u>Question 2D</u>. Is there a residually finite periodic FC-group G in which G/G′ is a direct product of cyclic groups, such that $G \notin \mathcal{SDF}$?

<u>Question 2E</u>. Is there a residually finite periodic FC-group G in which each abelian subgroup is a direct product of cyclic groups, such that $G \notin \mathcal{SDF}$?

These questions are both phrased to indicate my personal view that counterexamples will be found to exist. The main difficulty lies in proving that any given group is not in the class $\mathcal{SDF}$. In this connection there is one other question which we have to mention. In [41] , Gorčakov claimed to prove that a residually finite periodic FC-group with countable centre is an $\mathcal{SDF}$-group. His proof was incomplete and he withdrew the claim in his book [47] in which he also gave an example of a residually finite periodic FC-group with countable centre to show that the claim was actually false. Unfortunately, the argument that this group was not an $\mathcal{SDF}$-group was incorrect and I have been unable to determine whether or not the example is in the class $\mathcal{SDF}$. The following question, related to Question 2E, is therefore still outstanding.

Question 2F. Is there a residually finite periodic FC-group G with countable centre which is not an $\mathcal{SDF}$-group?

We observed earlier that Hall's Theorem and Gorčakov's Theorem correspond to theorems of Prüfer on countable abelian groups and abelian groups of finite exponent. The two theorems of Prüfer can be combined to give Kulikov's Criterion (Theorem 2.10) for an abelian group to be a direct product of finite cyclic groups. The theorems of Hall and Gorčakov can also be combined to give a similar criterion for a periodic FC-group to be in the class $\mathcal{SDF}$.

Theorem 2.30. (Gorčakov [45]) Let G be a periodic FC-group. Then G can be embedded in a direct product of finite groups if and only if G has an ascending chain of normal subgroups

$$1 = G_0 \leqslant G_1 \leqslant \ldots \leqslant \bigcup_{n=1}^{\infty} G_n = G,$$

such that, for each $x \in G_n - G_{n-1}$, there is a normal subgroup $N \geqslant G_{n-1}$ such that $x \notin N$ and G/N is finite of exponent dividing n!.

Proof. First, suppose $G \leqslant \mathrm{Dr}_{i\in I} F_i$. Let $G_n = G \cap \mathrm{Dr}\{F_i : |F_i| \leqslant n\}$; then the G_n clearly form an ascending chain of the required type.

Conversely, suppose G has an ascending chain of normal subgroups as in the statement of the theorem. Let

$$M_n = \bigcap \{N \triangleleft G : G/N \text{ is finite of exponent dividing } n! \text{ and } N \geqslant G_{n-1}\};$$

then $M_n \cap G_n = G_{n-1}$ so that $\bigcap_{n=1}^{\infty} M_n = 1$. In the embedding of G into $\prod_{n=1}^{\infty}(G/M_n)$ given by $x \mapsto (x M_n)$, each element of G has finite support so that G is isomorphic to a subgroup of $\mathrm{Dr}_{n=1}^{\infty}(G/M_n)$. Also G/M_n is residually finite of finite exponent and so $G/M_n \in \mathcal{SDF}$ (Corollary 2.29). Hence $G \in \mathcal{SDF}$. □

In Theorem 2.11, we used Kulikov's Criterion to prove that $Z_2(G)/Z(G)$

was a direct product of finite cyclic groups, obtaining a Kulikov chain in Z_2/Z from one in $[G,Z_2]$. The above result seems more difficult to apply than Kulikov's Criterion but a result of this type might be the way to approach Question 2A.

In Chapter 1, we gave our reasons for concentrating on periodic FC-groups, namely that G' and $G/Z(G)$ are periodic. However, there are questions concerning nonperiodic FC-groups which do not reduce to similar questions in the periodic case. Kurdačenko has written a series of papers concerning the question of when an FC-group can be embedded in the direct product of finite groups and a torsion-free abelian group. The following is perhaps his most striking result.

Theorem 2.31. (Kurdačenko [113]) Let T be a periodic FC-group. Every FC-group with periodic subgroup isomorphic to T is embeddable in a direct product of finite groups and a torsion-free abelian group if and only if $Z(T) = 1$.

Proof. Suppose first that $Z(T) = 1$ and let G have periodic subgroup P isomorphic to T. Let Z^* be the hypercentre of G; since $Z(P) = 1$ it follows that $Z^* \cap P = 1$. Now G/Z^* is a periodic FC-group with trivial centre and hence $G/Z^* \in \mathcal{SDF}$ (Corollary 2.26). Also G/P is a torsion-free abelian group and G is isomorphic to a subgroup of $(G/Z^*) \times (G/P)$.

Conversely, suppose that $Z(T) \neq 1$ and let z be an element of prime order p in $Z(T)$. Let $B = \mathrm{Dr}_{n=1}^{\infty} \langle b_n \rangle$ and $C = \mathrm{Dr}_{n=1}^{\infty} \langle c_n \rangle$ be free abelian groups of countable rank and let $A = \langle a \rangle$ be a group of order p. Consider the split extension U of $A \times B$ by C in which a is central, $[b_n, c_n] = a$ and $[b_m, c_n] = 1$, if $m \neq n$. Since U' is finite, U is an FC-group.

It is clear that $Z(U) = A \times B^p \times C^p$ and so $U/Z(U)$ is infinite. We show first that U can not be embedded in a direct product of finite groups and a torsion-free abelian group. If there were such an embedding, then it would have normal subgroups L and M such that $L \cap M = 1$, $U/L \in \mathcal{SDF}$ and U/M is torsion-free abelian. Then A, being the periodic subgroup of U, would be contained in M and hence $L \cong LA/A$ would be torsion-free. Since $U/L \in \mathcal{SDF}$, there is a normal subgroup K of finite index in U such that $K \cap LA = L$ and hence $K \cap A = 1$. But then K would be torsion-free and so $K \leq Z(U)$. This is contrary to $U/Z(U)$ being infinite.

Now let $W = T \ast U$, $R = \langle za^{-1} \rangle \leq Z(W)$ and let $G = W/R$. The periodic subgroup of W is $T \times A = T \times R$ and so the periodic subgroup of G is isomorphic to T. But G contains $UR/R \cong U/(U \cap R) \cong U$ and U is not embeddable in a direct product of finite groups and a torsion-free abelian group. □

3 Sections of direct products of finite groups

By a section of a group G we mean a homomorphic image of a subgroup of G. We saw in Corollary 2.25 that a residually finite periodic FC-group G is isomorphic to a section of a direct product of finite groups; that is $G \in \mathcal{QSDF}$. In particular, for any FC-group G, $G/Z(G) \in \mathcal{QSDF}$. Our main positive results in this Chapter will show that every countable periodic FC-group is a $\mathcal{QSDF}$-group and that, for any FC-group G, the derived subgroup G' is in the class $\mathcal{QSDF}$.

The class of countable periodic FC-groups was considered in Hall's important 1959 paper in which the main result for FC-groups depends on a construction which is valid in much greater generality.

Theorem 3.1. (Hall [49]) *Let the group G be generated by normal subgroups G_λ, $\lambda \in \Lambda$, and let the index set Λ be totally ordered. Let Ω be the subset of $\Lambda \times G$ consisting of pairs (λ, x) such that $x \in G_\lambda$. Let $\bar{G}$ be the group of permutations of Ω generated by the permutations $v_\mu(y)$, $\mu \in \Lambda$, $y \in G_\mu$, defined by*

$$v_\mu(y):(\lambda,x) \longmapsto \begin{cases} (\lambda, y^{-1}xy) & , \text{ if } \mu > \lambda, \\ (\lambda, xy) & , \text{ if } \mu = \lambda, \\ (\lambda, x) & , \text{ if } \mu < \lambda. \end{cases}$$

Then the mapping $\phi: v_\mu(y) \mapsto y$ determines a homomorphism of $\bar{G}$ onto G.

Proof. We need to show that if

$$v_{\mu_1}(y_1) \ldots v_{\mu_r}(y_r) = 1$$

then $y_1 \ldots y_r = 1$.

It is easily deduced from the definition of $v_\mu(y)$ that we have the following conjugation rules:

$$v_\nu(z)^{-1}\, v_\mu(y)\, v_\nu(z) = \begin{cases} v_\mu(z^{-1} yz) & , \text{ if } \nu \geq \mu, \\ v_\nu(z^{-1} yzy^{-1})\, v_\mu(y) & , \text{ if } \nu < \mu. \end{cases}$$

Therefore the element $v_{\mu_1}(y_1) \ldots v_{\mu_r}(y_r)$ of $\bar{G}$ can be written in the standard form $v_{\nu_1}(z_1) \ldots v_{\nu_s}(z_s)$, where $\nu_1 < \ldots < \nu_s$ and $z_1 z_2 \ldots z_s = y_1 \ldots y_r$ (and $s \leq r$). We may also assume that each z_i is non-trivial since $v_\nu(1)$ is the identity permutation. The permutation $v_{\nu_1}(z_1) \ldots v_{\nu_s}(z_s)$ maps (ν_s, x) to (ν_s, xz_s) contrary to it being the identity permutation. Thus, in its standard form, $v_{\nu_1}(z_1) \ldots v_{\nu_s}(z_s)$ is an empty product and so $y_1 \ldots y_r = z_1 \ldots z_s = 1$, as required. □

<u>Theorem 3.2.</u> (Hall [49]) <u>If G is a countable periodic FC-group then G is isomorphic to a factor group of a countable residually finite periodic FC-group.</u>

<u>Proof.</u> The countable group G can be expressed as the union of an ascending chain of finite normal subgroups $G = \bigcup_{n=1}^{\infty} G_n$. Taking these G_n to be the G_λ of Theorem 3.1 it is clear that $\bar{G}$ is countable, it being generated by the countable set of elements $\{v_n(x) : x \in G_n,\ n=1,2,\ldots\}$. Now let $\Omega_n = \{(n,x) : x \in G_n\}$ and let $M_n = \{g \in \bar{G} : (n,x)g = (n,x), \text{ for all } (n,x) \in \Omega_n\}$. Then $M_n \triangleleft \bar{G}$ and $\bar{G}/M_n$ is isomorphic to a group of permutations of the finite set Ω_n. Thus $\bar{G}/M_n$ is finite and clearly $\bigcap_{n=1}^{\infty} M_n = 1$, so that $\bar{G}$ is residually finite. Also $\bar{G}$ is the union of the finite subgroups $F_n = \langle v_m(x) : x \in G_m,\ m \leq n \rangle$ and the conjugation rules given above show that each F_n is a normal subgroup of $\bar{G}$. Hence $\bar{G}$ is a countable periodic FC-group. □

<u>Corollary 3.3.</u> <u>If G is a countable periodic FC-group, then</u> $G \in \mathcal{QSDF}$. □

In his original paper, Hall gave a more detailed discussion of the group $\bar{G}$ and in fact showed that a countable periodic FC-group G is in the class $\mathcal{QSD}(\mathcal{F} \cap \mathfrak{V}(G))$, where $\mathfrak{V}(G)$ is the variety generated by the group G.

<u>Corollary 3.4.</u> <u>If G is a periodic FC-group with G/Z(G) countable, then</u>

$G \in \mathcal{QSDF}$.

Proof. There is a countable normal subgroup N of G such that NZ = G. Thus G is isomorphic to the direct product of N and Z with the subgroup $N \cap Z$ amalgamated. By Corollary 3.3, $N \in \mathcal{QSDF}$ and Z, being a periodic abelian group, is isomorphic to a subgroup of a direct product of quasicyclic groups. Hence $Z \in \mathcal{SD}(\mathcal{QDF}) = \mathcal{QSDF}$. Therefore $G \in \mathcal{QD}(\mathcal{QSDF}) = \mathcal{QSDF}$. □

We know from Corollary 2.28 that, in any periodic FC-group $G, G'Z(G)/Z(G)$ is a direct product of countable groups X_i/Z, say. By the above result, each X_i is in the class $\mathcal{QSDF}$. If we also knew that $[X_i, X_j] = 1$, for each $i \neq j$, then we could say that $G'Z(G)$ was isomorphic to the direct product of the groups X_i with the subgroup Z amalgamated and hence $G'Z(G) \in \mathcal{QD}(\mathcal{QSDF}) =$ $= \mathcal{QSDF}$. The main problem then in showing that $G'Z(G)$ (and hence G') is a $\mathcal{QSDF}$-group is to construct the subgroups X_i with this property. It is convenient to begin with the following lemma which we give here in a form which will also be useful later.

Lemma 3.5. Let K be a normal subgroup of the periodic FC-group G containing $Z = Z(G)$. Then there is a normal subgroup $H \geq K$ such that $Z(H) = Z$ and $|H/K| \leq \max(\aleph_0, |Z(K)/Z|)$.

Proof. We construct normal subgroups K_n such that

$$K = K_0 \leq K_1 \leq \ldots \leq K_n \leq \ldots,$$

$|K_{n+1}/K_n| \leq \max(\aleph_0, |Z(K)/Z|)$ and $C_{K_n}(K_{n+1}) = Z$. Putting $H = \bigcup_{n=1}^{\infty} K_n$ will yield the required result.

Suppose then that K_{n-1} has been constructed. If $Y = Z(K_{n-1})$, then $Y \cap K = Z$ so that $|Y/Z| \leq |K_{n-1}/K| \leq \max(\aleph_0, |Z(K)/Z|)$. There is a set $\{1\} \cup \{x_i : i \in I\}$ of coset representatives of Z in Y with $|I| \leq |Y/Z|$. For each $i \in I$, we can choose an element $y_i \in G$ such that $[x_i, y_i] \neq 1$. It is clear that if $K_n = \langle K_{n-1}, y_i : i \in I\rangle^G$ then $C_{K_{n-1}}(K_n) = Z$ and $|K_n/K_{n-1}| \leq \max(\aleph_0, |Y/Z|) \leq \max(\aleph_0, |Z(K)/Z|)$. □

Theorem 3.6. (Tomkinson [105]) If G is a periodic FC-group then

$G' \in \mathcal{QSDF}$.

Proof. In fact, we prove by induction on $|G/Z(G)|$ that $G'Z(G) \in \mathcal{QSDF}$. By Corollary 3.4, we may assume that G/Z is uncountable, where $Z = Z(G)$.

The group G can be written as the union of an ascending chain of subgroups $G = \bigcup_{\alpha<\rho} G_\alpha$, where ρ is the least ordinal of cardinality $|G/Z|$ and, for each $\alpha<\rho$, $|G_\alpha| \leqslant \max(\aleph_0, |\alpha|)$.

Let $H_0 = Z$, $M_0 = G$ and suppose that for all ordinals $\beta < \alpha$ we have defined normal subgroups H_β, M_β such that

$$Z = H_0 < H_1 < \ldots < H_\beta < \ldots , \tag{1}$$

$$|H_\beta/Z| \leqslant \max(\aleph_0, |\beta|), \tag{2}$$

$$H_\beta \geqslant G_\beta, \tag{3}$$

$$Z(H_\beta) = Z , \tag{4}$$

$$M_\beta = C_G(H_\beta) , \tag{5}$$

$$\text{whenever } \beta\text{-1 exists, } H_\beta Z_{\beta-1} = G, \text{ where } Z_{\beta-1} = C_G(G/M_{\beta-1}) \tag{6}$$

$$\text{whenever } \beta \text{ is a limit ordinal, } H_\beta = \bigcup_{\gamma<\beta} H_\gamma . \tag{7}$$

We have a number of consequences of these conditions. Using (1) and (5), (4) and (5), and then (5) and (7), we obtain

$$G = M_0 \geqslant M_1 \geqslant \ldots \geqslant M_\beta \geqslant \ldots , \tag{8}$$

$$H_\beta \cap M_\beta = Z , \tag{9}$$

$$\text{whenever } \beta \text{ is a limit ordinal, } M_\beta = \bigcap_{\gamma<\beta} M_\gamma . \tag{10}$$

Also $M_\beta = C_G(H_\beta) = \bigcap \{C_G(F) : F$ is a finite normal subgroup of G contained in $H_\beta\}$. Now each $G/C_G(F)$ is finite and so G/M_β can be embedded in the cartesian product of the finite groups $G/C_G(F)$. There are at most $\max(\aleph_0, |\beta|)$ of these finite groups and so, by Lemma 2.22, we have

$$|G/Z_\beta| \leqslant \max(\aleph_0, |\beta|). \tag{11}$$

We construct normal subgroups H_α, M_α so that the conditions (1) to (7) are still satisfied.

If α is a limit ordinal we define $H_\alpha = \bigcup_{\beta<\alpha} H_\beta$; then all the conditions are satisfied. The only condition that needs checking is (2) and we have $|H_\alpha| = \Sigma_{\beta<\alpha} |H_\beta| \leqslant |\alpha| \max(\aleph_0, |\alpha|) \leqslant \max(\aleph_0, |\alpha|)$.

If α-1 exists, then $|G/Z_{\alpha-1}| \leqslant \max(\aleph_0, |\alpha-1|)$, by (11), and so there is a normal subgroup K of G such that $|K| \leqslant \max(\aleph_0, |\alpha|)$ and $KZ_{\alpha-1} = G$. It follows that $|H_{\alpha-1} G_\alpha K/Z| \leqslant |H_{\alpha-1}/Z||G_\alpha/Z||K| \leqslant \max(\aleph_0, |\alpha|)$. We use Lemma 3.5 to obtain a normal subgroup H_α of G such that $|H_\alpha/Z| \leqslant \max(\aleph_0, |\alpha|)$, $H_\alpha \geqslant H_{\alpha-1} G_\alpha K$ and $Z(H_\alpha) = Z$. It is immediate that conditions (1) to (7) are satisfied if we define $M_\alpha = C_G(H_\alpha)$. We can therefore define subgroups H_α, M_α, for all $\alpha > \rho$, to satisfy the conditions (1) to (11).

Now define $N_0 = M_1$ and, for $1 \leqslant \alpha < \rho$, $N_\alpha = [H_\alpha, G] M_{\alpha+1}$. Then, by applying Lemma 2.20 in the group G/Z, we have $\bigcap_{\alpha<\rho} N_\alpha = \bigcap_{\alpha<\rho} M_\alpha = Z$.

If $C_\alpha = C_G(G/N_\alpha)$, then $C_\alpha \geqslant Z_{\alpha+1}$, and so $H_{\alpha+2} C_\alpha \geqslant H_{\alpha+2} Z_{\alpha+1} = G$, by (6). Also $[H_\alpha, G] \leqslant [H_\beta, G] \leqslant N_\beta$, for all $\beta \geqslant \alpha$, and so the conditions of Lemma 2.21 are satisfied in the group G/Z. Hence G/Z is contained in the centrally restricted direct product of the groups $G/N_\alpha, \alpha < \rho$. Hence each element of G is contained in all but finitely many C_α. It follows that

each element of G' is contained in N_α for all but finitely many α. (12)

For each limit ordinal $\lambda < \rho$, let $\lambda' = \lambda + \omega$ be the succeeding limit ordinal. Since ρ is the least ordinal of some uncountable cardinal, we have $\lambda' < \rho$. For each limit ordinal $\lambda < \rho$, we define $X_\lambda = Z[H_{\lambda'}, G] \cap M_\lambda$. We show that $G'Z$ is the direct product of the groups $X_\lambda, \lambda < \rho$, and the group $Z[H_\omega, G]$ with the subgroup Z amalgamated.

We note first that these subgroups are mutually centralizing. For, if $\lambda < \mu$, then

$$[X_\lambda, X_\mu] \leqslant [ZH_{\lambda'} \cap M_\lambda, ZH_{\mu'} \cap M_\mu] \leqslant [H_{\lambda'}, M_\mu] \leqslant [H_\mu, M_\mu] = 1,$$

by condition (5). Also

$$[Z[H_\omega, G], X_\lambda] \leqslant [H_\omega, M_\lambda] \leqslant [H_\lambda, M_\lambda] = 1.$$

Secondly, we show that the product of the groups X_λ/Z and $Z[H_\omega, G]/Z$ is a direct product. It is clear that $\langle X_\lambda : \lambda < \rho \rangle \leqslant M_\omega$ and so $Z[H_\omega, G] \cap \langle X_\lambda : \lambda < \rho \rangle \leqslant Z[H_\omega, G] \cap M_\omega \leqslant Z(H_\omega \cap M_\omega) = Z$, by (10). Therefore $\langle Z[H_\omega, G]/Z, X_\lambda/Z : \lambda < \rho \rangle = Z[H_\omega, G]/Z \times \langle X_\lambda/Z : \lambda < \rho \rangle$.

In the subgroup $\langle X_\lambda : \lambda < \rho \rangle$, we have

$$\begin{aligned} X_\lambda \cap \langle X_\mu : \mu \neq \lambda \rangle &= X_\lambda \cap \langle X_\mu : \mu < \lambda \rangle \langle X_\mu : \mu > \lambda \rangle, \\ &\leqslant (H_{\lambda'} \cap M_\lambda) \cap \langle H_{\mu'} : \mu < \lambda \rangle M_{\lambda'}, \end{aligned}$$

$$= H_{\lambda'} \cap (M_\lambda \cap \langle H_{\mu'} : \mu < \lambda \rangle) M_{\lambda'},$$
$$\leq H_{\lambda'} \cap (M_\lambda \cap H_\lambda) M_{\lambda'},$$
$$= H_{\lambda'} \cap M_{\lambda'},$$
$$= Z.$$

Therefore $\langle X_\lambda/Z : \lambda<\rho\rangle = \mathrm{Dr}_{\lambda<\rho}(X_\lambda/Z)$.

It remains to prove that $G'Z$ is generated by the groups $Z[H_\omega,G]$ and $X_\lambda, \lambda<\rho$. It is clear that $Z \leq \langle Z[H_\omega,G], X_\lambda : \lambda<\rho\rangle \leq G'Z$ and so it is sufficient to show that $G' \leq \langle Z[H_\omega,G], X_\lambda : \lambda<\rho\rangle$. So let $x \in G' = \bigcup_{\alpha<\rho}[H_\alpha,G]$ and let λ be the least limit ordinal such that $x \in [H_\lambda,G]$; then λ can not be a limit of limit ordinals and so $\lambda = \omega$ or $\lambda = \lambda_o + \omega$, where λ_o is a limit ordinal.

If $\lambda = \omega$, then $x \in [H_\omega,G] \leq \langle Z[H_\omega,G], X_\lambda : \lambda < \rho\rangle$; so we suppose $\lambda = \lambda_o + \omega$. By (12), x is contained in all but finitely many $[H_\alpha,G]M_{\alpha+1}$. Considering the ordinals less than λ_o, we see that there is a $\delta < \lambda_o$ such that $x \in [H_\alpha,G]\,M_{\alpha+1}$, for all α such that $\delta \leq \alpha < \lambda_o$. For each such α we write $x = h_\alpha m_{\alpha+1}$, where $h_\alpha \in [H_\alpha,G]$ and $m_{\alpha+1} \in M_{\alpha+1}$. We prove by induction on α that, for all α such that $\delta \leq \alpha < \lambda_o$, there is a $z_\alpha \in Z$ such that $h_\alpha = h_\delta z_\alpha^{-1}$ and $m_{\alpha+1} = m_{\delta+1} z_\alpha$. We assume therefore that $h_\beta = h_\delta z_\beta^{-1}$ and $m_{\beta+1} = m_{\delta+1} z_\beta$, for all β such that $\delta \leq \beta < \alpha$. Thus $x = h_\delta m_{\delta+1}$, with $h_\delta \in [H_\delta,G]$ and $m_{\delta+1} \in \bigcap_{\beta<\alpha} M_{\beta+1} = M_\alpha$. But then $x = h_\delta m_{\delta+1} = h_\alpha m_{\alpha+1}$ gives

$$h_\alpha^{-1} h_\delta = m_{\alpha+1} m_{\delta+1}^{-1} \in H_\alpha \cap M_\alpha = Z$$

and so there is an element $z_\alpha \in Z$ such that $h_\alpha = h_\delta z_\alpha^{-1}$ and $m_{\alpha+1} = m_{\delta+1} z_\alpha$. Hence $h_\delta \in [H_\delta,G] \leq [H_{\lambda_o},G]$ and $m_{\delta+1} \in \bigcap_{\alpha<\lambda_o} M_{\alpha+1} = M_{\lambda_o}$ so that $x \in [H_{\lambda_o},G]M_{\lambda_o} \cap [H_\lambda,G] = [H_{\lambda_o},G]\,(M_{\lambda_o} \cap [H_\lambda,G]) = [H_{\lambda_o},G]X_{\lambda_o}$. By a further induction argument, we may assume that $[H_{\lambda_o},G] \leq \langle Z[H_\omega,G], X_\lambda : \lambda<\rho\rangle$ and so $x \in \langle Z[H_\omega,G], X_\lambda : \lambda<\rho\rangle$, as required.

We have now shown that $G'Z$ is the direct product of groups $Z[H_\omega,G]$ and $X_\lambda, \lambda<\rho$, with the subgroup Z amalgamated. For each $\lambda<\rho$, $|X_\lambda/Z|<|G/Z|$ and also $|Z[H_\omega,G]/Z|<|G/Z|$. Thus each factor is contained in a subgroup $L'Z$ where $|L/Z| < |G/Z|$ and, by Lemma 3.4, L can be chosen so that $Z(L) = Z$. By the induction hypothesis each $L'Z$ is in the class $\mathcal{QSDF}$. Hence $Z[H_\omega,G]$ and X_λ are in $\mathcal{QSDF}$ and so $G'Z \in \mathcal{QD}(\mathcal{QSDF}) = \mathcal{QSDF}$. □

In this proof we actually proved that $G'Z$ is the central product of groups X_λ with $|X_\lambda/Z| < |G/Z|$. In view of the fact that $G'Z/Z$ is known to be a direct product of countable groups it is reasonable to ask

Question 3A. If G is a periodic FC-group is $G'Z$ the central product of groups $X_i, i \in I$, such that $|X_i/Z|$ is countable?

We have already seen one example (2.9) of a periodic FC-group which is not in the class $\mathcal{QSDF}$ but it is possible to give a more systematic construction of such examples. We begin by introducing one of the properties of $\mathcal{QSDF}$-groups which makes this class so important and which we will use as a test for proving that a given group is not in the class $\mathcal{QSDF}$.

A group G is said to be a *$\mathcal{Z}$-group* if for each infinite cardinal $\mathfrak{m}$, and for each subset $S \subseteq G$ such that $|S| < \mathfrak{m}$, we have $|G:C_G(S)| < \mathfrak{m}$. In particular, taking $\mathfrak{m} = \aleph_0$, we see that every $\mathcal{Z}$-group is an FC-group.

Lemma 3.7. The class $\mathcal{Z}$ is $\mathcal{QSD}$-closed and hence $\mathcal{QSDF} \subseteq \mathcal{Z}$.

Proof. $\mathcal{Q}$-closure: Let $G \in \mathcal{Z}$ and $N \lhd G$. If S is a subset of G/N with $|S| < \mathfrak{m}$, then we can choose a set T of representatives of S (so that $S = \{tN: t \in T\}$) and $|T| = |S| < \mathfrak{m}$. Then $|G:C_G(T)| < \mathfrak{m}$ and, since $C_{G/N}(S) = C_{G/N}(TN/N) \geqslant C_G(T)N/N$, we have

$$|G/N: C_{G/N}(S)| = |C_G(T): N \cap C_G(T)| < \mathfrak{m}.$$

$\mathcal{S}$-closure: Let $G \in \mathcal{Z}$ and $H \leqslant G$. If S is a subset of H with $|S| < \mathfrak{m}$, then $|G:C_G(S)| < \mathfrak{m}$ and hence $|H:C_H(S)| = |H:H \cap C_G(S)| < \mathfrak{m}$.

$\mathcal{D}$-closure. Let $G = \mathrm{Dr}_{i\in I}\, G_i$ with $G_i \in \mathcal{Z}$. If S is a subset of G with $|S| < \mathfrak{m}$, then $J = \{i \in I: \pi_i(S) \neq 1\}$ has cardinality less than $\mathfrak{m}$. It is clear that $C_{G_i}(S) = C_{G_i}(\pi_i(S))$ and that $|\pi_i(S)| < \mathfrak{m}$ for each $i \in J$. If S is finite then J is finite and each $|G_i:C_{G_i}(S)|$ is finite and hence $|G:C_G(S)|$ is finite. If S is infinite, then $|J| \leqslant |S|$ and $|G_i:C_{G_i}(S)| \leqslant |S|$ so that $|G:C_G(S))| = \Sigma_{i\in J} |G_i:C_{G_i}(S)| \leqslant |S|^2 = |S| < \mathfrak{m}$. □

We now give two examples of an infinite extraspecial p-group G such that

$G \notin \mathfrak{Z}$. An *extraspecial p-group* is a group G such that $G' = Z(G)$, $|G'| = p$ and G/G' is an elementary abelian p-group. Since G' is finite, such a group is necessarily an FC-group.

Example 3.8. (Hall [49], Tomkinson [100]) Let $X = \mathrm{Dr}_{n=1}^{\infty} < x_n : x_n^p = 1 >$, $Y = \prod_{n=1}^{\infty} < y_n : y_n^p = 1 >$ and $Z = < z : z^p = 1 >$. We form the split extension G of $X \times Z$ by Y in which Z is central, $[x_n, y_n] = z$ and $[x_m, y_n] = 1$, if $m \neq n$. Then $Z(G) = G' = Z$ has order p and $G/G' \cong X \times Y$ is an elementary abelian p-group. It is clear that $C_G(X) = X \times Z$ so that $|G:C_G(X)| = \exp \aleph_0$ although $|X| = \aleph_0$. Hence $G \notin \mathfrak{Z}$ and so, by Lemma 3.7, $G \notin \mathcal{QSDF}$. □

Our previous example (2.9) of a group G which is not in the class $\mathcal{QSDF}$ contains the above extraspecial group as a subgroup, the subgroup generated by the socles of A and B, and so is not a $\mathfrak{Z}$-group. We note the obvious question about the possible extension of Corollary 2.8.

Question 3B. If G is a (periodic) $\mathfrak{Z}$-group, is $G/Z(G)$ necessarily in the class $\mathcal{SDF}$?

The second example, due to Ehrenfeucht and Faber, is a rather remarkable group in that although G' has order p and G/G' is an uncountable elementary abelian group, each abelian subgroup of G is countable. The one disadvantage of this construction is its reliance on GCH, the Generalized Continuum Hypothesis. Extraspecial groups play an important role in the remainder of this Chapter and so before giving the details of the Ehrenfeucht-Faber construction we outline the relationship between extraspecial p-groups and symplectic spaces over GF(p), the field of p elements.

First, let G be an extraspecial p-group and let z be a fixed generator of $G' = Z(G)$. The elementary abelian group G/G' can be written additively and considered as a vector space V over GF(p). If $x, y \in G$ then the commutator $[x,y]$ depends only on the cosets $\bar{x} = xG'$ and $\bar{y} = yG'$. Therefore we can define a mapping $\phi : V \times V \to GF(p)$ by $\phi(\bar{x}, \bar{y}) = k$, where $[x,y] = z^k$.

Since G is nilpotent of class two, $[x_1 x_2, y] = [x_1, y][x_2, y]$ and $[x, y_1 y_2] = [x, y_1][x, y_2]$. Hence the mapping ϕ is bilinear and also ϕ is an *alternate* mapping; i.e. $\phi(\bar{x}, \bar{x}) = 0$. It is a consequence of ϕ being alternate that it is also skew-symmetric; i.e. $\phi(\bar{y}, \bar{x}) = -\phi(\bar{x}, \bar{y})$. A vector space V over a field k equipped with an alternate bilinear map $\phi: V \times V \to k$ is called a *symplectic space*; V is said to be *non-degenerate* if, for each non-zero element $v \in V$, there is an element $w \in V$ such that $\phi(v,w) \neq 0$. So our definition of ϕ above associates with an extraspecial p-group G a non-degenerate symplectic space $V = G/G'$ over GF(p).

Much of the structure of an extraspecial p-group G may be deduced from the structure of its associated symplectic space; this is particularly so for finite or countably infinite extraspecial p-groups. We first introduce further notation for symplectic spaces. A subspace A of a symplectic space V is called *isotropic* if $\phi(v,w) = 0$, for all $v,w \in A$; this corresponds to an abelian subgroup of G. A *hyperbolic plane* is a two-dimensional non-degenerate symplectic space H; thus H has a basis $\{u,v\}$ such that $\phi(u,v) = 1$. A hyperbolic plane over GF(p) is the symplectic space associated with an extraspecial group of order p^3. If A and B are subspaces of V such that $\phi(a,b) = 0$ for all $a \in A$, $b \in B$ then we say that A and B are *orthogonal*. If, further, $A \cap B = 0$ then the subspace $A + B$ is called the *orthogonal sum* of A and B and written $A \oplus^{\perp} B$. An orthogonal sum of two subspaces of V corresponds to the direct product of two subgroups of G with the subgroup Z amalgamated. The *orthogonal complement* of a subspace A is the subspace $A^{\perp} = \{v \in V : \phi(a,v) = 0, \text{ for all } a \in A\}$. The orthogonal complement of a subspace in V corresponds to the centralizer of a subgroup in G.

Theorem 3.9. A non-degenerate symplectic space V of countable dimension is an orthogonal sum of hyperbolic planes.

Proof. We note first that if the subspace U is a finite orthogonal sum of hyperbolic planes, then $V = U \oplus U^{\perp}$. Since V is non-degenerate we must have $U \cap U^{\perp} = 0$ and so we need only show that $U + U^{\perp} = V$. Suppose that $\bigoplus^{\perp n}_{i=1} H_i$, where H_i has a basis $\{u_i,v_i\}$ with $\phi(u_i,v_i) = 1$. If x is any element of V, then $\phi(u_i,x) = s_i$, $\phi(v_i,x) = t_i$, say. It is easily checked that

$$x - \sum_{i=1}^{n} s_i v_i + \sum_{i=1}^{n} t_i u_i \in U^{\perp}$$

and so $x \in U \oplus U^{\perp}$.

Now let V have a basis $\{x_1, x_2, \ldots\}$; we construct hyperbolic planes H_n

such that $\langle x_1,\ldots,x_n\rangle \subseteq \bigoplus_{i=1}^{\perp n} H_i$ and hence $V = \bigoplus_{i=1}^{\infty} H_i$. Suppose we have obtained $H_1,\ldots,H_{n-1}$ and let $U = \bigoplus_{i=1}^{\perp n} H_i$. By the above, $V = U \oplus U^{\perp}$ and we can choose an element $u_n \in U^{\perp}$ such that $x_n \in \langle U,u_n\rangle$. Since V is non-degenerate, there is an element $v_n \in U^{\perp}$ such that $\phi(u_n,v_n) = 1$ and we can define $H_n = \langle u_n,v_n\rangle$. □

Corollary 3.10. A countably infinite extraspecial p-group G is the direct product with amalgamated centre of groups of order p^3. Hence $G \in \mathcal{QDF}$.

If p = 2, then G is the direct product with amalgamated centre of groups each isomorphic to D, the dihedral group of order 8.

If p > 2, then either G has exponent p and is the direct product with amalgamated centre of groups each isomorphic to the unique nonabelian group of order p^3 and exponent p or G has exponent p^2 and is the direct product with amalgamated centre of a nonabelian group of order p^3 and exponent p^2 and a countable group of exponent p.

Proof. The first part of the corollary follows immediately from Theorem 3.9. A nonabelian group of order 8 is either a dihedral group D or a quaternion group Q. So a countable extraspecial 2-group is a direct product of groups D_i and Q_j with amalgamated centre. Suppose there are infinitely many factors D_i, then $D_1D_2 \cong Q_1Q_2$, $D_3D_4 \cong Q_3Q_4$, etc. ([55],p.355). Thus G is a product of groups each isomorphic to Q. But using $Q_1Q_2 \cong D_1D_2$ again, we then see that G is a product of groups each isomorphic to D.

If p > 2, then letting X be the nonabelian group of order p^3 and exponent p and Y the nonabelian group of order p^3 and exponent p^2 we can use a similar argument to that above making use of the fact that $Y_1Y_2 \cong Y_1X_2$. □

The argument used in proving Theorem 3.9 depended on the subspace $U = \bigoplus_{i=1}^{n-1} H_i$ having finite dimension and so can not be extended to spaces of uncountable dimension. Further constructions of extraspecial groups with bad properties will now depend on constructing symplectic spaces of uncountable dimension with bad properties. We first need to state more precisely how we shall construct an extraspecial group from a given symplectic space.

Let V be a non-degenerate symplectic space over GF(p) with a basis

$\{v_i : i \in I\}$. We define the group $G(V)$ to be the group generated by elements z and $x_i, i \in I$, subject to the relations

$$z^p = x_i^p = [x_i, z] = 1, \quad [x_i, x_j] = z^{\phi(v_i, v_j)} .$$

It is immediate from these relations that $G' = \langle z \rangle$. To see that $G(V)$ is in fact an extraspecial group with G' having order p we must show that the defining relations do not imply that z is trivial. This can be done by giving an explicit construction of a nonabelian group satisfying the relations.

Let F be the field of p elements and define a total ordering $<$ on the index set I. The set $F \times V$ can be made into a group $G(V, <)$ by defining

$$(\alpha, \Sigma\, \alpha_i v_i)(\beta, \Sigma\, \beta_i v_i) = (\alpha + \beta + \Sigma_{i<j}\, \alpha_i \beta_j\, \phi(v_i, v_j),\ \Sigma(\alpha_i + \beta_i) v_i).$$

It is straightforward to check that this multiplication does, in fact, determine a group and that, if $i < j$,

$$[(0, v_i), (0, v_j)] = (\phi(v_i, v_j), 0)$$
and
$$[(0, v_j), (0, v_i)] = (-\phi(v_i, v_j), 0).$$

Therefore $G(V)$ is nonabelian and, since V is non-degenerate, $Z(G) = G' = \langle z \rangle$ has order p. We also have $G/G' = \mathrm{Dr}_{i \in I} \langle \bar{x}_i \rangle$ so that $G(V)$ is an extraspecial p-group and V is its associated symplectic space. If $p > 2$, then $G(V)$ has exponent p. The construction could, of course be easily varied to give a group of exponent p^2 by simply changing one of the relations $x_0^p = 1$ to $x_0^p = z$. This variation will not be needed here.

A maximal abelian subgroup of $G(V)$ will contain $G' = Z(G)$ and so will correspond to a maximal isotropic subspace of V. The construction of the Ehrenfeucht-Faber group therefore is essentially the construction of a symplectic space V of uncountable dimension in which each isotropic subspace has countable dimension. In fact the construction can be used with higher cardinalities.

Let V be a vector space of dimension $\mathfrak{m}^+$ over $GF(p)$, where $\mathfrak{m}$ is some infinite cardinal and $\mathfrak{m}^+$ is the successor cardinal, and let V have basis $\{v_\varepsilon : \varepsilon < \kappa\}$, where κ is the least ordinal of cardinality $\mathfrak{m}^+$. We let

$V_\varepsilon = \langle v_\alpha : \alpha < \varepsilon \rangle$ and define an alternate bilinear form on V inductively.

Lemma 3.11. For each alternate bilinear mapping $\phi : V_\varepsilon \times V_\varepsilon \to GF(p)$ with $|\varepsilon| = \mathfrak{m}$ and each family $\mathcal{R}$ of $\mathfrak{m}$-dimensional subspaces of V_ε with $|\mathcal{R}| \leq \mathfrak{m}$, there exists an extension $\phi' : V_{\varepsilon+1} \times V_{\varepsilon+1} \to GF(p)$ of ϕ such that ϕ' is an alternate bilinear mapping and, for each pair $(v,U) \in V_\varepsilon \times \mathcal{R}$, there is an element $u \in U$ such that $\phi'(u, v + v_\varepsilon) \neq 0$.

Proof. We may well-order the set $V_\varepsilon \times \mathcal{R}$ with order type μ, where μ is the least ordinal of cardinality $\mathfrak{m}$, and denote the λth element of $V_\varepsilon \times \mathcal{R}$ by $(v(\lambda), U(\lambda))$. We define inductively a linearly independent set of elements $\{u_\alpha : \alpha < \mu\}$ with $u_\alpha \in U(\alpha)$, for each $\alpha < \mu$. Let u_0 be any non-zero element of $U(0)$. If the elements $u_\beta, \beta < \alpha$, have been defined so that the set $\{u_\beta : \beta < \alpha\}$ is linearly independent then this set has cardinality less than $\mathfrak{m}$ and, since $\dim U(\alpha) = \mathfrak{m}$, we can choose an element $u_\alpha \in U(\alpha)$ such that $\{u_\beta : \beta \leq \alpha\}$ is linearly independent. Thus the linearly independent set $\{u_\alpha : \alpha < \mu\}$ can be defined inductively and can be extended to form a basis $\mathcal{B}$ of V_ε; $\mathcal{B} \cup \{v_\varepsilon\}$ will then be a basis for $V_{\varepsilon+1}$.

To extend ϕ to $V_{\varepsilon+1} \times V_{\varepsilon+1}$, we need to define $\phi'(b, v_\varepsilon)$ for each $b \in \mathcal{B}$; the values of $\phi'(v_\varepsilon, v_\varepsilon)$ and $\phi'(v_\varepsilon, b)$ then follow from ϕ' being alternate. We define $\phi'(u_\alpha, v_\varepsilon)$ to be different from $-\phi(u_\alpha, v(\alpha))$. If $b \in \mathcal{B} - \{u_\alpha : \alpha < \mu\}$ then we can define $\phi'(b, v_\varepsilon)$ arbitrarily.

If $(v,U) \in V_\varepsilon \times \mathcal{R}$, then there is some $\alpha < \mu$ such that $(v,U) = (v(\alpha), U(\alpha))$ so that $\phi'(u_\alpha, v(\alpha) + v_\varepsilon) = \phi(u_\alpha, v(\alpha)) + \phi'(u_\alpha, v_\varepsilon) \neq 0$ and $u_\alpha \in U$ is the required element. □

Theorem 3.12. (Ehrenfeucht and Faber [20]) Under the assumption $\exp \mathfrak{m} = \mathfrak{m}^+$,

(i) there is a symplectic space V over GF(p) of dimension $\mathfrak{m}^+$ in which each isotropic subspace has dimension at most $\mathfrak{m}$.

(ii) there is an extraspecial p-group G(V) of cardinality $\mathfrak{m}^+$ in which each abelian subgroup has cardinality at most $\mathfrak{m}$.

Proof. Using $\exp \mathfrak{m} = \mathfrak{m}^+$, the set of $\mathfrak{m}$-dimensional subspaces of V has cardinality $\mathfrak{m}^+$ and so we can label the $\mathfrak{m}$-dimensional subspaces of V as $U_\varepsilon, \mu \leq \varepsilon < \kappa$, in such a way that U_ε is a subspace of V_ε.

Let $\phi_\mu : V_\mu \times V_\mu \to GF(p)$ be any alternate bilinear mapping. Extend ϕ_μ

inductively as follows. Suppose $\phi_\varepsilon : V_\varepsilon \times V_\varepsilon \to GF(p)$ is an alternate bilinear mapping extending ϕ_μ; then ϕ_ε can be extended to $\phi_{\varepsilon+1} : V_{\varepsilon+1} \times V_{\varepsilon+1} \to GF(p)$ using Lemma 3.11 with $\mathcal{R} = \{U_\alpha : \alpha \leq \varepsilon\}$. If ε is a limit ordinal and ϕ_δ has been defined for each $\delta < \varepsilon$, we define ϕ_ε by $\phi_\varepsilon(v,u) = \phi_\delta(v,u)$, where $(v,u) \in V_\delta \times V_\delta \subseteq \bigcup_{\delta<\varepsilon} V_\delta \times V_\delta = V_\varepsilon \times V_\varepsilon$. Let ϕ be the resulting extension of ϕ_μ to $V \times V$.

Suppose that U is an isotropic subspace of V having dimension $\mathfrak{m}^+$. Then U contains an $\mathfrak{m}$-dimensional subspace U_ε, say. Since $U \not\subseteq V_\varepsilon$, U contains an element of the form $v = \Sigma_{\alpha<\delta} n_\alpha v_\alpha + v_\delta$, where $\delta > \varepsilon$. The mapping $\phi_{\delta+1}$ was obtained from ϕ_δ by applying Lemma 3.11 with $\mathcal{R} = \{U_\alpha : \alpha < \delta\}$. Thus $U_\varepsilon \in \mathcal{R}$ and, since $w = \Sigma_{\alpha<\delta} n_\alpha v_\alpha \in V_\delta$, there is an element $u \in U_\varepsilon \subseteq U$ such that $\phi(u, w + v_\delta) \neq 0$. That is, $\phi(u,v) \neq 0$ contrary to U being isotropic. This contradiction shows that V has no isotropic subspace of dimension $\mathfrak{m}^+$.□

It is clear that this example of Ehrenfeucht and Faber is also outside the class $\mathfrak{Z}$. For, a maximal abelian subgroup coincides with its centralizer and so for each maximal abelian subgroup A of G, we have $|A| = \mathfrak{m}$ but $|G:C_G(A)| = |G:A| = \mathfrak{m}^+$.

The main disadvantage of this example is its reliance on GCH. In their Remark 1, Ehrenfeucht and Faber [20] indicate reasons for believing that it may be possible to avoid this.

<u>Question 3C</u>. Can one construct an extraspecial p-group G of cardinality $\mathfrak{m}^+$ (or exp $\mathfrak{m}$) in which each abelian subgroup has cardinality at most $\mathfrak{m}$, without the use of GCH?

The next theorem will show that the property of an infinite extraspecial group E having all its maximal abelian subgroups of the same cardinality as E is related to a condition which is very close to E being a $\mathfrak{Z}$-group. We define a *$\mathfrak{Y}$-group* to be a locally finite group G in which, for each infinite cardinal $\mathfrak{m}$ and for each subgroup H with $|H| < \mathfrak{m}$, we have $|G:N_G(H)| < \mathfrak{m}$. In particular, consideration of $\mathfrak{m} = \aleph_0$ shows that every $\mathfrak{Y}$-group is a periodic FC-group and it is clear that every periodic $\mathfrak{Z}$-group is a $\mathfrak{Y}$-group. We therefore have

$$\mathcal{QSDF} \subseteq \{\text{periodic } \mathfrak{Z}\text{-groups}\} \subseteq \mathfrak{Y} \subseteq \{\text{periodic FC-groups}\}.$$

As for $\mathfrak{Z}$-groups, we can prove some closure properties of the class $\mathfrak{Y}$.

Lemma 3.13. The class $\mathfrak{Y}$ is QS-closed.

Proof. Q-closure Let $G \in \mathfrak{Y}$ and $N \triangleleft G$. If H/N is a subgroup of G/N with $|H/N| < \mathfrak{m}$ then there is a subgroup U of G such that $|U| < \mathfrak{m}$ and $H = UN$. Now $N_G(H) \geq N_G(U)$ and so

$$|G/N:N_{G/N}(H/N)| = |G:N_G(H)| \leq |G:N_G(U)| < \mathfrak{m}$$

and so $G/N \in \mathfrak{Y}$.

S-closure. Let $G \in \mathfrak{Y}$ and $U \leq G$. If H is a subgroup of U such that $|H| < \mathfrak{m}$ then $|G:N_G(H)| < \mathfrak{m}$ and hence $|U:N_U(H)| = |U:U \cap N_G(H)| < \mathfrak{m}$ so that $U \in \mathfrak{Y}$. □

However, we can not prove that $\mathfrak{Y}$ is D-closed. For, suppose G is a $\mathfrak{Y}$-group that is not a $\mathfrak{Z}$-group and let H be an infinite subgroup of G such that $|G:C_G(H)| > |H|$. In the direct product $G \times G$ consider $K = \{(h,h) : h \in H\}$, the diagonal subgroup of $H \times H$. Then certainly $|K| = |H|$ but if x,y are in different cosets of $C_G(H)$, then $x^{-1} K x = \{(x^{-1}hx,h):h \in H\}$ and $y^{-1} K y = \{(y^{-1}hy,h):h \in H\}$ are distinct conjugates of K. Hence K has $|G:C_G(H)|$ distinct conjugates in $G \times G$ and so $G \times G \notin \mathfrak{Y}$.

Theorem 3.14. (Tomkinson [103]) An extraspecial group E is a $\mathfrak{Y}$-group if and only if, for each infinite subgroup H of E and each maximal abelian subgroup A of H, $|A| = |H|$.

Proof. Let E be an extraspecial $\mathfrak{Y}$-group and let A be a maximal abelian subgroup of E, then $C_E(A) = A$. If $|A| < |E|$, then $|E:C_E(A)| = |E:A| > |A|$. Since A has finite exponent it is a direct product of finite abelian groups and so may be expressed in the form $B \times Y$, where Y is a finite subgroup containing $Z = Z(E)$. Thus $|B| = |A|$, $C_E(A) = C_E(B) \cap C_E(Y)$ and $|E:C_E(Y)|$ is finite so that $|E:C_E(B)| = |E:C_E(A)| > |A| = |B|$. But $B \cap E' = 1$ and so $N_E(B) = C_E(B)$ and hence $|E:N_E(B)| > |B|$ contrary to E being a $\mathfrak{Y}$-group. Therefore $|A| = |E|$.

Now let H be any infinite subgroup of E and A a maximal abelian subgroup of H. If $H \cap Z = 1$, then H is abelian so that H = A and hence $|A| = |H|$. So we may assume that $H \geq Z$. By Lemma 3.5, there is a subgroup H_1 of E containing H such that $Z(H_1) = Z$ and $|H_1/H| \leq \max(\aleph_0, |Z(H)/Z|)$. Since $Z(H_1) = Z$, H_1 is an extraspecial group and, as a subgroup of the $\mathfrak{Y}$-group E, H_1 is also a $\mathfrak{Y}$-group, by Lemma 3.13. There is a maximal abelian subgroup A_1 of H_1 containing A and, since $A_1 \cap H = A$, we have $|A_1/A| \leq |H_1/H| \leq$ $\leq \max(\aleph_0, |Z(H)|) \leq |A|$ so that $|A_1| = |A|^2 = |A|$. By the first paragraph of the proof, $|A_1| = |H_1|$ and so $|A| = |A_1| = |H_1| = |H|$, as required.

Conversely, let E be an infinite extraspecial p-group which is not a $\mathfrak{Y}$-group. Then E has an infinite subgroup U such that $|E:N_E(U)| > |U|$. Clearly $U \not\geq Z$ and so $U \cap Z = 1$. That is, $U \cap E' = 1$ and so $N_E(U) = C_E(U) \geq UZ$ and also U is abelian. Since E/UZ is elementary abelian, $C_E(U)/UZ$ is a direct factor of E/UZ and so there is a subgroup H such that $HC_E(U) = E$ and $H \cap C_E(U) = UZ$. Now let A be a maximal abelian subgroup of H containing UZ. Then $A \leq H \cap C_E(U) = UZ$ and so $A = UZ$. Therefore $|A| = |U|$, but $|H/A| = |H/UZ| = |E:C_E(U)| > |U|$ so that $|A| < |H|$, as required. □

The two examples of extraspecial groups that we have considered (Example 3.8 and Theorem 3.12) are therefore periodic FC-groups which are not in the class $\mathfrak{Y}$. The following theorem shows that these examples are typical in that periodic FC-groups which are not in $\mathfrak{Y}$ must involve an extraspecial group which is not in $\mathfrak{Y}$.

Theorem 3.15. (Tomkinson [103]) *Let G be a periodic FC-group. Then G is a $\mathfrak{Y}$-group if and only if each extraspecial section of G is a $\mathfrak{Y}$-group.*

Proof. By the QS-closure of the class $\mathfrak{Y}$ given in Lemma 3.13, it is clear that if G is a $\mathfrak{Y}$-group then each section of G is a $\mathfrak{Y}$-group. We assume therefore that G is not in the class $\mathfrak{Y}$ and show that G has an extraspecial section which is not a $\mathfrak{Y}$-group. The method is to repeatedly replace G by a subgroup or factor group thus imposing more and more restrictions on G until we are left with an extraspecial group.

Since G is not in the class $\mathfrak{Y}$, it has an infinite subgroup U such that $|G:N_G(U)| > |U|$. Clearly by factoring out if necessary we may assume that

$$U_G = 1\ , \tag{1}$$

where U_G denotes the core of U.

Writing Z for Z(G) we rely heavily on the fact that G/Z, being residually finite, is a $\mathcal{QSDF}$-group (Corollary 2.25) and hence, by Lemma 3.7, is a $\mathcal{Z}$-group. Therefore $|G:C_G(UZ/Z)| \leqslant |UZ/Z| \leqslant |U|$ and so $|C_G(UZ/Z):C_G(UZ/Z) \cap N_G(U)| > |U|$. Replacing G by its subgroup $UC_G(UZ/Z)$, we may assume:

$$UZ \lhd G \ \underline{\text{and}} \ \ |C:N_C(U)| > |U|\ , \text{ where } C = C_G(UZ/Z). \tag{2}$$

Again we may factor out U_G so that we may assume both (1) and (2). If $c \in C$, then $F_c = [U,c]$ is a finite subgroup of Z and $c \in C_G(UF_c/F_c)$. Also $C_G(UF_c/F_c) \leqslant C$ and so $C = \bigcup_{c\in C} C_G(UF_c/F_c)$. Each subgroup F_c is contained in U^G and $|U^G| = |U|$. Therefore there are at most $|U|$ distinct subgroups F_c and so there is one such subgroup for which $|C_G(UF_c/F_c):C_G(UF_c/F_c) \cap N_G(U)| > |U|$. We choose F to be a finite subgroup of $U^G \cap Z$ of minimal order subject to $|C_G(UF/F):C_G(UF/F) \cap N_G(U)| > |U|$. Since $F \leqslant Z$ and $U_G = 1$, we have $U \cap F = 1$.

We now replace G by its subgroup $UC_G(UF/F)$ and write C for $C_G(UF/F)$. This gives

$$UF \lhd G,\ F \leqslant Z,\ C_G(UF/F) = C, |C:N_C(U)| > |U|,\ U \cap F = 1 \tag{3}$$

Again we may factor out U_G so that we may assume (1) and (3). There is a subgroup $D \leqslant F$ such that F/D is cyclic of prime order p, say. By the minimality of F, $|C_G(UD/D):C_G(UD/D) \cap N_G(U)| \leqslant |U|$. It is clear that $C_G(UD/D) \leqslant N_G(UD) \cap C$. Conversely, let $x \in N_C(UD)$; then $[x,UD] \leqslant UD \cap Z = (U \cap Z)D = D$ and so $x \in C_G(UD/D)$. Therefore $C_G(UD/D) = N_C(UD)$ and we have $|N_C(UD):N_C(UD) \cap N_G(U)| \leqslant |U|$. Since $|C:N_C(U)| > |U|$, it follows that $|C:N_C(UD)| > |U|$.

We therefore replace G by G/D, U by UD/D, etc, to obtain

$$\underline{\text{F has order}}\ p,\ F \leqslant Z,\ UF \lhd G,\ C = C_G(UF/F),\ G = UC, \quad |C:N_C(U)| > |U| \ \underline{\text{and}}\ U_G = 1. \tag{4}$$

Let $u \in U$, $c \in C$; then $[c,u] \in F$ and so $[c,u^p] = [c,u]^p = 1$ and also $[c,[u_1,u_2]] = [[c,u_1],[c,u_2]] = 1$ so that C centralizes $U'U^p$. In particular $U'U^p \triangleleft U C = G$ and so $U'U^p = 1$. That is, U is an elementary abelian p-group. It follows that $U \leq C$ and so $G = C$. Summing up, we have

G <u>has a central subgroup</u> F <u>of order</u> p <u>and an infinite elementary abelian</u> p-<u>subgroup</u> U <u>such that</u> $G = C_G(UF/F)$, $U_G = 1$ <u>and</u> $|G:N_G(U)| > |U|$. (5)

Let N be a normal subgroup of G maximal with respect to $N \cap UF = 1$. It is clear that $N_G(U) \leq N_G(UN)$; conversely let $x \in N_G(UN)$, then $x^{-1}Ux \leq UN \cap UF = U(N \cap UF) = U$ and so $x \in N_G(U)$. Hence $N_G(UN) = N_G(U)$ and we have $|G/N: N_{G/N}(UN/N)| = |G:N_G(U)| > |U| = |UN/N|$. If $X \triangleleft G$ and $X > N$, then $X \cap UF \neq 1$ and so there is a minimal normal subgroup M of G contained in $X \cap UF$. If $M \neq F$, then $[M,G] \leq M \cap F = 1$ and so $M \leq Z$; but then we would have $MF \leq Z$ and since $U \cap Z = 1$, we have $MF = UF \cap MF \leq UF \cap Z = (U \cap Z)F = F$. Hence $M = F$ and so every normal subgroup of G properly containing N must contain F.

Replacing G by G/N, U by UN/N, F by FN/N, we have in addition to (5)

F <u>is the unique minimal normal subgroup of</u> G <u>and, in particular</u> Z <u>is a cyclic or quasicyclic p-group</u>. (6)

If $x \in N_G(U)$, then $[x,U] \leq U \cap F$, since $G = C_G(UF/F)$, and so $[x,U] = 1$. Therefore

$$N_G(U) = C_G(U) = C_G(UF) \triangleleft G \qquad (7)$$

If $u \in U$ and $x,y \in G$, then $[x,u] \in F$ so that $[x^p,u] = [x,u]^p = 1$ and $[[x,y],u] = [[x,u],[y,u]] = 1$. Thus $x^p \in N_G(U)$ and $[x,y] \in N_G(U)$. That is

$G/N_G(U)$ <u>is an elementary abelian</u> p-<u>group</u> (8)

We now construct an ascending chain of subgroups

$$UF = A_o < A_1 < \ldots < A_\alpha < \ldots \quad (\alpha < \rho),$$

where ρ is the least ordinal of cardinality $|U|$, such that

(i) A_α/F is an abelian p-group and $|A_\alpha| = |U|$,

(ii) $A_\alpha \cap Z = F$,

(iii) $A_\alpha N_G(U) < A_{\alpha+1}N_G(U)$,

(iv) $A_\beta = \bigcup_{\alpha<\beta} A_\alpha$, for limit ordinals $\beta < \rho$.

Then, letting $A = \bigcup_{\alpha<\rho} A_\alpha$, we see that A/F is an abelian p-group $A \cap Z = F$ and $|A:N_A(U)| > |U|$.

We construct the subgroups A_α inductively. The limit ordinal case is clear as (i) and (ii) follow immediately from (iv). So we may assume that $A_{\alpha-1}$ has been constructed. Let $C_{\alpha-1} = C_G(A_{\alpha-1}/F)$; then since $A_{\alpha-1}/F \cong A_{\alpha-1}Z/Z$ and $G/Z \in \mathfrak{Z}$, we have $|G/C_{\alpha-1}| \leqslant |A_{\alpha-1}| = |U|$. Therefore $|C_{\alpha-1}:C_{\alpha-1} \cap N_G(U)| > |A_{\alpha-1}|$ and so $C_{\alpha-1} > C_{\alpha-1} \cap A_{\alpha-1}N_G(U)$.

Let $\langle \bar{z} \rangle$ be the unique subgroup of $A_{\alpha-1}Z/A_{\alpha-1}$ of order p and let a_1 be a p-element in $C_{\alpha-1} - A_{\alpha-1}N_G(U)$. (We can choose a_1 to be a p-element since $C_{\alpha-1}/A_{\alpha-1}N_G(U)$ is a p-group). Then $\langle \bar{z}, \bar{a}_1 \rangle$ is a finite abelian p-subgroup of $C_{\alpha-1}/A_{\alpha-1}$ and so $|C_{\alpha-1}:C_{C_{\alpha-1}}(\langle \bar{z}, \bar{a}_1 \rangle)|$ is finite. Therefore there is a p-element $a_2 \in C_{C_{\alpha-1}}(\langle \bar{z}, \bar{a}_1 \rangle) - A_{\alpha-1}N_G(U)$. Thus $\langle \bar{z}, \bar{a}_1, \bar{a}_2 \rangle$ is again a finite abelian p-subgroup of $C_{\alpha-1}/A_{\alpha-1}$. Also $\langle a_1, a_2 \rangle A_{\alpha-1}N_G(U)/A_{\alpha-1}N_G(U)$ is not cyclic and so $\langle \bar{a}_1, \bar{a}_2 \rangle$ is not cyclic. It follows that $\langle \bar{z}, \bar{a}_1, \bar{a}_2 \rangle$ is a noncyclic group and so contains an element $\bar{a}$, say, no power of which is equal to $\bar{z}$. That is, there is a p-element $a \in C_{\alpha-1} - A_{\alpha-1}N_G(U)$ such that $\langle \bar{a} \rangle$ has trivial intersection with $A_{\alpha-1}Z/A_{\alpha-1}$.

Let $A_\alpha = \langle A_{\alpha-1}, a \rangle$. Clearly A_α/F is an abelian p-group, $|A_\alpha| = |A_{\alpha-1}| = |U|$ and $A_\alpha N_G(U) > A_{\alpha-1}N_G(U)$. Also

$$A_\alpha \cap Z = \langle A_{\alpha-1}, a \rangle \cap A_{\alpha-1} Z \cap Z = A_{\alpha-1} \cap Z = F.$$

This completes our inductive definition of the subgroups A_α and, replacing G by $A = \bigcup_{\alpha<\rho} A_\alpha$, we now have

<u>G is a p-group</u>, G/F <u>is abelian</u>, $|F| = p$, U <u>is an elementary abelian subgroup of</u> G <u>and</u> $|G:N_G(U)| > |U|$ (9)

If $x,y \in G$, then $[x,y] \in F$ and so $[x^p,y] = [x,y]^p = 1$ and so $x^p \in Z$. Therefore G/Z is elementary abelian. As in the argument leading to (6) we can factor out a normal subgroup N which is maximal with respect to $N \cap UF = 1$ and so we may assume (9) together with

$$G/Z \text{ \underline{is elementary abelian and} } Z \text{ \underline{is cyclic or quasicyclic}.} \tag{10}$$

Now $UF \cap Z = (U \cap Z)F = F$ and so there is a subgroup $X \geqslant UF$ such that $X \cap Z = F$ and G/X is countable (or finite). Then $|X:N_X(U)| > |U|$. Also $X/F \cong XZ/Z$ is elementary abelian and so, replacing G by X, we have

G <u>is a p-group such that</u> $|G'| = p$ <u>and</u> G/G' <u>is elementary abelian</u>, U <u>is a subgroup such that</u> $U \cap G' = 1$ <u>and</u> $|G:N_G(U)| > |U|$. (11)

If $Z > G'$, then there is a subgroup $Y \geqslant UG'$ such that $G/G' = (Z/G') \times (Y/G')$ and hence $Z(Y) = Z \cap Y = G'$. Thus Y is an extraspecial group which is not a $\mathfrak{Y}$-group. □

This proof relied heavily on the fact that G/Z is a $\mathfrak{Z}$-group rather than just a $\mathfrak{Y}$-group but we have not been able to adapt the proof to give the corresponding result for $\mathfrak{Z}$-groups.

<u>Question 3D</u>. If G is a periodic FC-group in which each extraspecial section is a $\mathfrak{Z}$-group, is G itself a $\mathfrak{Z}$-group?

<u>Question 3E</u>. If G is a periodic FC-group in which each extraspecial section is a $\mathcal{QSDF}$-group, is G itself a $\mathcal{QSDF}$-group?

The main problem in this area is that of determining the relationship between the three classes being considered. At present we do not even know that the classes are distinct, all the examples that we have constructed of periodic FC-groups which are not $\mathcal{QSDF}$-groups lie outside the class $\mathfrak{Y}$. The discussion following Lemma 3.13 then does not necessarily show that $\mathfrak{Y}$ is not $\mathcal{D}$-closed; it shows that if $\mathfrak{Y} \not\subseteq \mathfrak{Z}$, then $\mathfrak{Y}$ is not $\mathcal{D}$-closed.

<u>Question 3F</u>. Are the three classes $\mathcal{QSDF}$, {periodic $\mathfrak{Z}$-groups} and $\mathfrak{Y}$

distinct?

A first step towards answering this question might be a more detailed investigation of infinite extraspecial groups. Some results and examples were given in [103] and we give one of these examples here as it seems likely that this is a $\mathfrak{Z}$-group which is not in the class $\mathcal{QSDF}$. If, in fact, the group is in the class $\mathcal{QSDF}$ then it shows that extraspecial $\mathcal{QSDF}$-groups can be surprisingly complicated and, in particular, the following question would have a negative answer.

<u>Question 3G.</u> Can an extraspecial $\mathcal{QSDF}$-group be embedded in a direct product of groups of order p^3 with amalgamated centre?

<u>Example 3.16.</u> (Tomkinson [103]) There is an extraspecial $\mathfrak{Z}$-group of cardinality $\aleph_1$ which can not be embedded in a direct product of groups of order p^3 with amalgamated centre.

The group is obtained as G(V) where V is a symplectic space over GF(p) which can not be embedded in an orthogonal sum of hyperbolic planes. For G(V) to be a $\mathfrak{Z}$-group, we require the space V to satisfy the condition:

For each subspace U of V having countable dimension, dim $(V/U^{\perp})$ is countable.

We let V have a basis consisting of x_i, y_i and z_λ, where i takes all ordinal values less than ω_1 and λ takes all <u>limit</u> ordinal values less than ω_1. We define an alternating map $\phi: V \times V \to GF(p)$ by

$$\psi(x_i, x_j) = \psi(y_i, y_j) = \psi(z_\lambda, z_\mu) = 0,$$
$$\phi(x_i, y_i) = 1,\ \phi(x_i, y_j) = 0, \text{ if } i \neq j,$$
$$\phi(y_i, z_\lambda) = 0,$$
$$\phi(x_i, z_\lambda) = 1, \text{ if } i < \lambda,\ \phi(x_i, z_\lambda) = 0 \text{ if } i \geqslant \lambda.$$

For each ordinal $\varepsilon < \omega_1$, let $V_\varepsilon = \langle x_i, y_i, z_\lambda : i < \varepsilon, \lambda \leqslant \varepsilon \rangle$; then $\bigcup_{\varepsilon<\omega_1} V_\varepsilon$. Each subspace U of countable dimension is contained in some V_ε and to show that dim $(V/U^{\perp})$ is countable it is clearly sufficient to show that $V = V_\varepsilon \oplus V_\varepsilon^\perp$. Certainly $V_\varepsilon^\perp \supseteq \langle x_j, y_j : j \geq \varepsilon \rangle$ and if $\varepsilon = \lambda+n$, where λ is a limit ordinal and $1 \leq n < \omega$, then

$$z_\lambda + y_\lambda + y_{\lambda+1} + \dots + y_{\lambda+n-1} - z_\mu \in V_\varepsilon^\perp, \text{ for all } \mu > \varepsilon,$$

so that $z_\mu \in V_\varepsilon \oplus V_\varepsilon^\perp$. If ε is a limit ordinal, then $z_\varepsilon - z_\mu \in V_\varepsilon^\perp$ for all $\mu > \varepsilon$, so that again $z_\mu \in V_\varepsilon \oplus V_\varepsilon^\perp$. Therefore $V_\varepsilon \oplus V_\varepsilon^\perp = V$, as required.

Now suppose that $V \subseteq H = \bigoplus^\perp_{i<\omega_1} H_i$, where H_i is a hyperbolic plane with basis $\{u_i, v_i\}$ such that $\phi(u_i, v_i) = 1$. For each $i < \omega_1$, we write $K_i = \bigoplus^\perp_{j<i} H_j$.

For each $i < \omega_1$, there is a least ordinal $j(i)$ such that $i < j(i) < \omega_1$ and $V_i \subseteq K_{j(i)}$. For each $j < \omega_1$, there is a least ordinal $i(j)$ such that $j < i(j) < \omega_1$ and $K_j \cap V \subseteq V_{i(j)}$. Given an ordinal $i < \omega_1$, we define

$$i_o = i, \quad j_o = j(i),$$
$$i_n = i(j_{n-1}), \; j_n = j(i_{n-1}), \text{ for all integers } n \geq 1.$$

Then the sequences $\{i_n\}$ and $\{j_n\}$ are strictly increasing and

$$\bigcup_{n=1}^\infty K_{j_n} \cap V = \bigcup_{n=1}^\infty V_{i_n}$$

Let $\lambda = \ell ub\,\{j_n\}$ and $\mu = \ell ub\,\{i_n\}$; then λ and μ are limit ordinals and

$$K_\lambda \cap V = \bigcup_{i<\mu} V_i . \tag{*}$$

We shall call a limit ordinal μ for which there exists a limit ordinal $\lambda = \lambda(\mu)$ satisfying (*) a μ-*ordinal*. We have shown that if i is any ordinal less than ω_1, then there is a μ-ordinal μ such that $i < \mu < \omega_1$. In particular, there are uncountably many μ-ordinals.

If μ is a μ-ordinal and $\lambda = \lambda(\mu)$, then since $H = K_\lambda \oplus K_\lambda^\perp$, there is an element $k_\lambda \in K_\lambda$ such that $k_\lambda - z_\mu \in K_\lambda^\perp$. Now suppose, if possible, that there is no element $k \in H$ such that $k - z_\mu \in K^\perp_{\lambda(\mu)}$ for uncountably many μ-ordinals μ. Then, for each $\lambda < \omega_1$, there is a smallest μ-ordinal $\mu(\lambda)$ such that, for each element k in the countable set K_λ and for each μ-ordinal $\nu \geq \mu(\lambda)$, $k - z_\nu \notin K^\perp_{\lambda(\nu)}$.

Choose some μ-ordinal μ_o and define $\lambda_o = \lambda(\mu_o)$ and, for each integer $n \geq 1$, define $\mu_n = \mu(\lambda_{n-1})$ and $\lambda_n = \lambda(\mu_{n-1})$. Let $\mu = \ell ub\{\mu_n\}$ and $\lambda = \ell ub\{\lambda_n\}$; then μ and λ are limit ordinals and

$$K_\lambda \cap V = \bigcup_{n=1}^{\infty} (K_{\lambda_n} \cap V) = \bigcup_{n=1}^{\infty} (\bigcup_{i<\mu_n} V_i) = \bigcup_{i<\mu} V_i.$$

Thus μ is a μ-ordinal and $\lambda = \lambda(\mu)$. If $k \in K_\lambda$, then $k \in K_{\lambda_n}$ for some n and so $k - z_\nu \notin K^\perp_{\lambda(\nu)}$ for any μ-ordinal $\nu \geq \mu_{n+1} = \mu(\lambda_n)$. In particular, there is no $k \in K_\lambda$ such that $k - z_\mu \in K^\perp_{\lambda(\mu)}$. This contradiction shows that there is some element $k \in H$ such that $k - z_\mu \in K^\perp_{\lambda(\mu)}$, for uncountably many μ-ordinals μ .

It follows that

$$\phi(k,y_i) = \phi(k,z_\lambda) = 0, \quad \text{for all } i,\lambda < \omega_1 ,$$
$$\phi(x_i,k) = 1, \quad \text{for all } i < \omega_1.$$

We now consider the subspace $\bar{V} = V + \langle k \rangle$ of H. Letting $\bar{V}_\varepsilon = V_\varepsilon + \langle k \rangle$, we clearly have $\bar{V} = \bigcup_{\varepsilon<\omega_1} \bar{V}_\varepsilon$. This allows us to repeat the arguments above, defining a $\bar{\mu}$-ordinal to be a limit ordinal μ for which there exists a limit ordinal $\lambda = \bar{\lambda}(\mu)$ such that

$$K_\lambda \cap \bar{V} = \bigcup_{i<\mu} \bar{V}_i .$$

We are then able to show that there is an element $h \in H$ such that $h - z_\mu \in K^{\perp}_{\bar{\lambda}(\mu)}$, for uncountably many $\bar{\mu}$-ordinals μ. It follows that

$$\phi(h,x_i) = \phi(h,y_i) = \phi(h,z_\lambda) = 0, \text{ for all } i,\lambda, < \omega_1,$$
$$\phi(h,k) = 1.$$

Suppose that $h \in K_\delta$ and let μ be a μ-ordinal such that $\lambda(\mu) > \delta$ and $k - z_\mu \in K^\perp_{\lambda(\mu)}$. But $(h,k - z_\mu) = 1$ contrary to $h \in K_\delta \subseteq K_{\lambda(\mu)}$. Thus V can not be embedded in H, an orthogonal sum of hyperbolic planes. □

One of the main problems in constructing examples in this area is that of proving that a given group is not in the class $\mathcal{QSDF}$. There seems to be no simple test for this and it is, in fact, the reason for introducing the classes $\mathfrak{Z}$ and $\mathfrak{Y}$ at this point. There was a similar problem in Chapter 2 for showing that a given residually finite periodic FC-group is not in the class $\mathcal{SDF}$. There the problem was largely overcome by considering a slightly different question and obtaining the characterization of residually finite periodic FC-groups in Theorem 2.24. It is quite possible that we are making a mistake in concentrating here on $\mathcal{QSDF}$-groups and there may be a related construction which gives a characterization of all periodic FC-groups. S.E. Stonehewer has suggested that it may be possible simply to replace $\mathcal{F}$ by some class of groups which contains both $\mathcal{F}$ and extraspecial groups. We have no information on the following specific question.

Question 3H. Is there a periodic FC-group which is not in the class $\mathcal{QSD}(\mathcal{FA})$? ($\mathcal{FA}$ denotes the class of finite-by-abelian groups; i.e., groups with finite derived subgroups).

The idea behind the consideration of extraspecial groups was the feeling that they are relatively simple objects. There is one other result which indicates how wrong this feeling was. Felgner considered groups of the form

$$G(I) = \langle z, x_i\,(i \in I) : z^p = x_i^p = [x_i, z] = 1,\ [x_i, x_j] = z \text{ if } i < j \rangle,$$

where I is a totally ordered set of cardinality $\mathfrak{m}$. Felgner showed that for $p \neq 2$, such groups satisfy the conditions for the application of Shelah's theory of saturated models and non-isomorphism theorems [91] and so was able to prove

Theorem 3.17. (Felgner [31]) If p is an odd prime and $\mathfrak{m}$ an uncountable cardinal then there are exp $\mathfrak{m}$ non-isomorphic extraspecial p-groups of cardinality $\mathfrak{m}$ of the form G(I). □

This result, being based on Shelah's theory, is purely an existence theorem. Again one feels that extraspecial groups should be more accessible than the general theories covered by Shelah's work and that a more constructive proof should be possible.

Question 3I. Can one construct exp $\mathfrak{m}$ non-isomorphic extraspecial p-groups of cardinality $\mathfrak{m}$? Can one give any proof to include the case $p = 2$?

It is also worth noting that the groups $G(I)$ all contain an abelian subgroup of cardinality $\mathfrak{m}$. For, we can find a set S of pairs of elements $(i_1, i_2) \in I \times I$ with $i_1 < i_2$ such that $|S| = \mathfrak{m}$ and if (j_1, j_2) is another pair in S then either $i_2 < j_1$ or $j_2 < i_1$. Then $< x_{i_1} x_{i_2}^{-1} : (i_1, i_2) \in S >$ is an abelian subgroup of cardinality $\mathfrak{m}$.

However, in the construction of the Ehrenfeucht-Faber group in Theorem 3.12, there seems to be considerable freedom in extending the alternating map $\phi: V_\varepsilon \times V_\varepsilon \to GF(p)$ to a mapping $\phi': V_{\varepsilon+1} \times V_{\varepsilon+1} \to GF(p)$. This suggests that many different Ehrenfeucht-Faber groups could be constructed and we ask

Question 3J. If $\mathfrak{m}$ is a regular cardinal, are there exp $\mathfrak{m}$ non-isomorphic extraspecial p-groups of cardinality $\mathfrak{m}$ in which each abelian subgroup has cardinality less than $\mathfrak{m}$?

4 Inverse limits of finite groups, locally inner automorphisms

A *directed set* I is a set on which is defined a partial order $<$ satisfying the additional condition:

(DS) if $i, j \in I$, then there is a $k \in I$ such that $i < k$ and $j < k$.

Let $\{G_i : i \in I\}$ be a family of finite groups G_i indexed by the directed set I. Suppose that, for each pair $i, j \in I$ with $i < j$, there is a homomorphism $\theta_{ji} : G_j \to G_i$ and these homomorphisms satisfy the conditions

(IS1) θ_{ii} is the identity mapping on G_i

(IS2) if $i < j < k$, then $\theta_{kj}\, \theta_{ji} = \theta_{ki} : G_k \to G_i$.

The family $\{G_i : i \in I\}$ together with the homomorphisms θ_{ji} forms an *inverse system* of finite groups.

We form the cartesian product $\Pi_{i\in I} G_i$ and consider the subgroup L consisting of all $(x_i) \in \Pi_{i\in I} G_i$ such that, whenever $i < j$, $x_j\, \theta_{ji} = x_i$. Then L is the *inverse limit* of the system $\{G_i\}$ and is usually written $L = \varprojlim G_i$. Although the homomorphisms θ_{ji} are a necessary part of the description of the inverse system $\{G_i\}$ they are usually suppressed in the notation and this should not lead to any confusion. If $\pi_i : L \to G_i$ is the natural projection map then it is clear that, whenever $i < j$, $\pi_j\, \theta_{ji} = \pi_i$. As the projection maps are combined with other maps in this Chapter it seems better to write them on the right, contrary to the convention we used in Chapter 2.

We have defined the inverse limit $L = \varprojlim G_i$ as a subgroup of the abstract group $\Pi_{i\in I} G_i$ but many of the properties of L that we require are most easily understood by introducing a topology on L. The finite groups G_i can be considered as topological groups with the discrete topology and then we take the cartesian product $\Pi_{i\in I} G_i$ to have the product topology. In this product topology, $\Pi_{i\in I} G_i$ has a basis consisting of the sets $\Pi_{i\in I} U_i$, where U_i is a subset of G_i and, for all but finitely many $i \in I$, U_i is the whole of G_i. The open sets in $\Pi_{i\in I} G_i$ are just unions of sets in this basis.

The subgroup L of $\Pi_{i\in I} G_i$ has the induced topology and becomes a topological group, the inverse limit of the system $\{G_i\}$ of finite groups with the discrete topology. Such an inverse limit is called a *profinite group*.

If $\pi_i: L \to G_i$ is the natural projection map, then we let $K_i = \text{Ker}\, \pi_i = L \cap \Pi_{j \neq i} G_j$. It is clear that K_i is a normal open subgroup of finite index in L and is also closed, its complement being the union of cosets each of which is open. All the open sets and the closed subgroups are easily described in terms of these subgroups K_i and these descriptions help us to keep the topological ideas to a minimum.

Lemma 4.1. (i) Each open subset of L containing the identity contains one of the subgroups K_i.

(ii) The open subsets of L are precisely the unions of cosets of subgroups K_i.

Proof. (i) Let U be an open set containing the identity in L. Then since L carries the induced topology, $U = L \cap V$, where V is an open subset of $\Pi_{i \in I} G_i$ containing the identity. By the definition of the product topology, V contains a subset $\Pi_{i \in I} U_i$, where U_i is a subset of G_i containing the identity and for all but finitely many $i \in I$, U_i is equal to G_i. Let J be the finite set $\{i \in I : U_i \neq G_i\}$; then $U = L \cap V \supseteq L \cap \Pi_{i \notin J} G_i = \bigcap_{i \in J} K_i$. Since J is finite, it follows from (DS) that there is a $j \in I$ such that $i \prec j$, for all $i \in J$. Since $\pi_j \theta_{ji} = \pi_i$, it follows that $K_j = \text{Ker}\, \pi_j \leqslant \text{Ker}\, \pi_i = K_i$, for all $i \in J$, and so $K_j \leqslant \bigcap_{i \in J} K_i \subseteq U$.

(ii) Since K_i is open so is any coset of K_i and hence all unions of such cosets are open. Conversely, let U be any open subset of L and let $x \in U$. Then $U x^{-1}$ is an open subset containing the identity and so, by (i), there is an $i = i(x) \in I$ such that $K_i \subseteq Ux^{-1}$ and hence $x \in K_i x \subseteq U$. It follows that $U = \bigcup_{x \in U} K_{i(x)} x$. □

Lemma 4.2. (i) If H is a subgroup of L then the closure of H is the subgroup $\bar{H} = \bigcap_{i \in I} H K_i$.

(ii) Any finite subset of L is closed.

Proof. (i) Each subgroup $H K_i$ is open since it is a union of cosets of K_i, and so each $H K_i$ is also closed. Therefore $\bigcap_{i \in I} H K_i$ is a closed subgroup of L. To show that $\bigcap_{i \in I} H K_i$ is the closure of H we show that every open subset U such that $U \cap \bigcap_{i \in I} H K_i \neq \emptyset$ also intersects H non-trivially.

Let $x \in U \cap \bigcap_{i\in I} HK_i$; then $x^{-1}U \supseteq K_i$, for some $i \in I$, and so $U \supseteq xK_i$. But $x \in HK_i$ and so $x = hk$, for some $h \in H$, $k \in K_i$. Therefore $xK_i = hK_i \subseteq U$. In particular, $h \in U$ and so $U \cap H \neq \emptyset$.

(ii) Since $\bigcap_{i\in I} K_i = 1$, part (i) shows that $\{1\}$ is closed. Therefore any singleton is closed and hence any finite subset of L is closed. □

It is a fairly simple matter to characterize profinite groups completely by their topological properties (e.g. [54] p.52). However, in line with our policy of keeping the topology to a minimum we give only the most important part of that result.

Theorem 4.3. A profinite group is compact.

Proof. Let $L = \varprojlim G_i \leq \Pi_{i\in I} G_i$. By Tihonov's Theorem, $\Pi_{i\in I} G_i$ is compact. The subgroups $T_i = \Pi_{j\neq i} G_i$ are open in $\Pi_{i\in I} G_i$. If $x = (x_i)$ is any element of $\Pi_{i\in I} G_i$ outside L, then there is a pair $i \prec j$ such that $x_j\theta_{ji} \neq x_i$. The set of elements with ith component x_i and jth component x_j is a coset of $T_i \cap T_j$ and so is open. This set contains x but no element of L. Thus the complement of L is a union of these open sets and hence L is closed. Therefore L is a closed subset of the compact group $\Pi_{i\in I} G_i$ and so is itself compact. □

The compactness property will usually be used to say that if $\{C_\lambda : \lambda \in \Lambda\}$ is a family of closed subsets in L with the *finite intersection property* (that is, each finite intersection $C_{\lambda_1} \cap \ldots \cap C_{\lambda_n}$ is non-empty) then $\bigcap_{\lambda\in\Lambda} C_\lambda \neq \emptyset$.

In the discussion above we concentrated on the kernels K_i. These clearly form a residual system of normal subgroups of finite index (see Chapter 1). Any residually finite group G has such a residual system $\{N_i : i \in I\}$ and most applications of profinite groups to abstract group theory involve the construction of a profinite group from a given residually finite group G with a suitably chosen residual system $\{N_i : i \in I\}$ of normal subgroups of finite index.

The index set I becomes a directed set if we define $i \prec j$ to mean $N_i \geq N_j$. The family $\{G/N_i : i \in I\}$ of finite groups becomes an inverse system if, whenever $N_i \geq N_j$, we define θ_{ji} to be the natural homomorphism from G/N_j to to G/N_i. We form the inverse limit $L = \varprojlim (G/N_i)$.

The mapping $\theta: x \mapsto (x N_i)$ is a monomorphism from G into L; we show that $G\theta$ is a dense subgroup of L. Let $x = (x_i N_i) \in L$. Given a fixed $j \in I$, define $y = (y_i N_i)$, where $y_i = x_j^{-1} x_i$, for each $i \in I$. Then $y_j N_j = 1N_j$ and so $y \in K_j = \mathrm{Ker}\ \pi_j$. But $(x_j N_i) = x_j\theta \in G\theta$ so that $x = (x_j N_i)(y_i N_i) \in (G\theta) K_j$. Therefore $(G\theta) K_j = L$, for each $j \in I$ and so the closure of $G\theta$ is L, by Lemma 4.2. Thus G is isomorphic to a dense subgroup of the compact group L. We say that L is the *profinite completion* of G with respect to the residual system $\{N_i : i \in I\}$ of normal subgroups of finite index.

Different residual systems may give rise to different profinite completions. For example, if G is the infinite cyclic group and N_i is the subgroup of index p^i then the profinite completion of G with respect to $\{N_i : i=1,2,\ldots\}$ is the group of p-adic integers Z_p. If we take the residual system consisting of all subgroups of finite index then the completion obtained is isomorphic to the cartesian product of the Z_p's.

For our purposes, a rather more typical example would be a direct product of finite groups $G = \mathrm{Dr}_{n=1}^{\infty} G_n$ with a residual system consisting of the subgroups $N_i = \mathrm{Dr}_{n=i+1}^{\infty} G_n$. Then the completion of G is just the cartesian product $\prod_{n=1}^{\infty} G_n$.

We do not give a detailed discussion of the relationships between different completions of G. [For periodic FC-groups some information is given by Hartley [52]]. However it will be useful to introduce an equivalence between residual systems. This is most easily achieved by first considering residual systems satisfying an additional condition. A residual system $\mathcal{N} = \{N_i : i \in I\}$ of normal subgroups of finite index in G is called a *residual filter* if

(RF) whenever $M \geq N$, $M \triangleleft G$ and $N \in \mathcal{N}$, we have $M \in \mathcal{N}$.

It is clear that the intersection of all residual filters containing a given residual system $\mathcal{N}$ is again a residual filter and we call this the *residual filter generated by* $\mathcal{N}$.

Lemma 4.4. (i) If $\mathcal{N}$ is a residual filter and $N_1, \ldots, N_k \in \mathcal{N}$, then $N_1 \cap \ldots \cap N_k \in \mathcal{N}$.

(ii) The residual filter generated by the residual system $\mathcal{N}$ consists of all normal subgroups of G which contain an element of $\mathcal{N}$.

Proof. (i) Since $\mathcal{N}$ is a residual system, there is an $N \in \mathcal{N}$ such that $N \leq N_1 \cap \ldots \cap N_k$ (RS2) and so, by (RF), $N_1 \cap \ldots \cap N_k \in \mathcal{N}$.

(ii) Certainly any residual filter containing $\mathcal{N}$ must contain all normal subgroups of G which contain elements of $\mathcal{N}$. But these subgroups do form a residual filter. They obviously satisfy conditions (RF) and (RS1). Suppose $M_1, \ldots, M_k$ are normal subgroups of G containing the subgroups $N_1, \ldots, N_k$ from $\mathcal{N}$. There is a subgroup $N \in \mathcal{N}$ such that $N \leq N_1 \cap \ldots \cap N_k$ and hence $M_1 \cap \ldots \cap M_k$ contains an element of $\mathcal{N}$. □

We shall say that two residual systems of G are *equivalent* if they generate the same residual filter.

Lemma 4.5. Let $\mathcal{M} = \{M_i : i \in I\}$ and $\mathcal{N} = \{N_j : j \in J\}$ be two residual systems of the group G.

(i) $\mathcal{M}$ and $\mathcal{N}$ are equivalent if and only if, for each $i \in I$, there is a $j = j(i) \in J$ such that $N_j \leq M_i$ and, for each $j \in J$, there is an $i = i(j) \in I$ such that $M_i \leq N_j$.

(ii) If $\mathcal{M}$ and $\mathcal{N}$ are equivalent, then the profinite completions of G with respect to $\mathcal{M}$ and $\mathcal{N}$ are topologically isomorphic.

Proof. (i) is straightforward, using Lemma 4.4 (ii).

(ii) It is sufficient to consider the case in which $\mathcal{M}$ is the residual filter generated by $\mathcal{N}$ so that $\mathcal{N} \subseteq \mathcal{M}$. Let $L(\mathcal{M})$ be the profinite completion of G with respect to $\mathcal{M}$ so that $L(\mathcal{M}) \leq \prod_{M \in \mathcal{M}} (G/M)$ and similarly for $L(\mathcal{N})$. Define $\phi : L(\mathcal{M}) \to L(\mathcal{N})$ by $(x_M M)\phi = (x_M M : M \in \mathcal{N})$. It is clear that ϕ is a surjective homomorphism. Suppose $(x_M M) \in \operatorname{Ker} \phi$; then for all $N \in \mathcal{N}$, $x_N N = N$. But for each $M \in \mathcal{M}$, there is an $N \in \mathcal{N}$ such that $N \leq M$ and $x_M M = (x_N N)\pi_{NM} = x_N M = M$. So $\operatorname{Ker} \phi = 1$ and ϕ is an isomorphism. Also if $K_N = \operatorname{Ker}(\pi_N : L(\mathcal{N}) \to G/N)$, then $K_N \phi^{-1} = \operatorname{Ker}(\pi_N : L(\mathcal{M}) \to G/N)$ and it follows from Lemma 4.1 that ϕ is continuous. □

Having defined an embedding θ of the residually finite group G into a profinite completion L we make the usual identification of G and $G\theta$ and we can then give G the induced topology. This topology can be described entirely in terms of the subgroups N_i in the residual system $\mathcal{N}$.

Lemma 4.6. Let L be the profinite completion of the residually finite group G with respect to the residual system $N = \{N_i : i \in I\}$ and, for each $i \in I$, let $K_i = \text{Ker}(\pi_i : L \to G/N_i)$. Then

(i) $N_i = G \cap K_i$,

(ii) $\bar{N}_i = K_i$,

(iii) the open subsets of G are the unions of cosets of subgroups N_i,

(iv) if H is a subgroup of G, then the closure of H in G is $\bigcap_{i \in I} H N_i$.

Proof. (i) Let $x \in G$; then $x\pi_i = x N_i \in G/N_i$. Thus $x \in G \cap K_i$ if and only if $x \in N_i$.

(ii) By Lemma 4.2, $\bar{N}_i = \bigcap_{j \in I} N_i K_j = \bigcap_{j \in I} (G \cap K_i) K_j = \bigcap_{j > i} (G \cap K_i) K_j = \bigcap_{j > i} G K_j \cap K_i$. But we have already seen that $G K_j = L$, for all $j \in I$, and so $\bar{N}_i = \bigcap_{j > i} G K_j \cap K_i = K_i$.

(iii) Certainly each subgroup N_i is open, by (i), and hence each coset of N_i is open. Therefore all unions of cosets of subgroups N_i are open. Conversely, let T be an open subset of G; then $T = G \cap U$, where U is an open subset of L. By Lemma 4.1 (ii), U has the form $U = \bigcup_{\lambda \in \Lambda} K_\lambda x_\lambda$ and since $L = K_\lambda G$ each x_λ can be taken to be an element of G. Thus $T = G \cap \bigcup_{\lambda \in \Lambda} K_\lambda x_\lambda = \bigcup_{\lambda \in \Lambda} (G \cap K_\lambda x_\lambda) = \bigcup_{\lambda \in \Lambda} (G \cap K_\lambda) x_\lambda = \bigcup_{\lambda \in \Lambda} N_\lambda x_\lambda$, as required.

(iv) Let $H \leq G$; then the closure of H in G is equal to the closure of H in L intersected with G. By Lemma 4.2, this is $\bigcap_{i \in I} H K_i \cap G = \bigcap_{i \in I} (H K_i \cap G) = \bigcap_{i \in I} H(K_i \cap G) = \bigcap_{i \in I} H N_i$. □

Lemma 4.7. Let S be a subset and H a closed subgroup of the residually finite group G with the topology induced from a profinite completion. Then $C_G(S)$ and $N_G(H)$ are closed subgroups of G.

Proof. (i) Let $C = C_G(S)$ and consider the closure $\bigcap_{i \in I} C N_i$ of C in G. Now $[\bigcap_{i \in I} C N_i, S] \leq \bigcap_{i \in I} [C N_i, S] \leq \bigcap_{i \in I} N_i = 1$ and so $\bigcap_{i \in I} C N_i = C$ and C is closed.

(ii) Let $N = N_G(H)$ and consider the closure $\bigcap_{i \in I} N N_i$ of N in G. Now $[\bigcap_{i \in I} N N_i, H] \leq \bigcap_{i \in I} [N N_i, H] \leq \bigcap_{i \in I} H N_i = H$ and so $\bigcap_{i \in I} N N_i = N$ and N is closed.□

Many of our results will depend purely on cardinality arguments and it will be necessary for us to know the cardinality of a profinite completion and the index of certain subgroups. As far as I am aware these results have not previously appeared in the literature; the proof given is based on an argument shown to me by D. Holt who had used it to give a direct proof of Theorem 4.14. All our results will follow from Theorem 4.9 which gives the index of a closed subgroup H of a profinite group L and we begin with some simple observations about the cosets of H.

Lemma 4.8. Let L be a profinite group with residual system $\{K_i : i \in I\}$ forming a basis of neighbourhoods of the identity and let H be a closed subgroup of L.

(i) The right cosets of H in L are intersections of the form $\bigcap_{i\in I} H K_i x_i$ such that $H K_i x_j = H K_i x_i$, whenever $K_i \geq K_j$.

(ii) The cosets $\bigcap_{i\in I} H K_i x_i$ and $\bigcap_{i\in I} H K_i y_i$ are equal if and only if $H K_i x_i = H K_i y_i$, for all $i \in I$.

Proof. (i) Each right coset of H is of the form $Hx = (\bigcap_{i\in I} H K_i)x = \bigcap_{i\in I} H K_i x$, which obviously satisfies the condition stated.

Conversely, each coset $H K_i x_i$ is closed and the family of these cosets has the finite intersection property. For, if $H K_1 x_1, \ldots, H K_n x_n$ is a finite subfamily, there is a $K_s \leq K_1 \cap \ldots \cap K_n$ and so $H K_s x_s \subseteq H K_1 x_1 \cap \ldots \cap H K_n x_n$. Since L is compact, the intersection $\bigcap_{i\in I} H K_i x_i$ is non-empty and contains some element x, say. Now $x \in H K_i x_i$ and so $H K_i x_i = H K_i x$ and hence $\bigcap_{i\in I} H K_i x_i = \bigcap_{i\in I} H K_i x = (\bigcap_{i\in I} H K_i)x = Hx$.

(ii) If $\bigcap_{i\in I} H K_i x_i = \bigcap_{i\in I} H K_i y_i = Hx$, then we have $x \in H K_i x_i$ and $x \in H K_i y_i$ so that $H K_i x_i = H K_i x = H K_i y_i$. The converse is clear.□

Theorem 4.9. Let L be a profinite group with the residual system $\{K_i : i \in I\}$ forming a system of neighbourhoods of the identity. Let H be a closed subgroup of infinite index in L and let the set of distinct subgroups $H K_i$ be indexed by a set J. Then $|L:H| = \exp |J|$.

Proof. To simplify notation, we write H_j for the subgroup HK_j, $j \in J$. The index set J may be well-ordered and so we may assume that $J = \{j : j < \rho\}$, where ρ is the least ordinal of cardinality $|J|$. A second well-ordering is defined on J by defining a map $\phi : \{\alpha : \alpha < \rho\} \to J$ as follows:

Let $\phi(1)$ be any element of J. Let α be any ordinal less than ρ and suppose that we have defined $\phi(\beta)$ for each $\beta < \alpha$.

(I) If $\{j : H_j \geq H_{\phi(\beta_1)} \cap \ldots \cap H_{\phi(\beta_k)}$, for some $\beta_1, \ldots, \beta_k < \alpha\} \not\subseteq \{\phi(\beta) : \beta < \alpha\}$, then define $\phi(\alpha)$ to be the smallest $j \notin \{\phi(\beta) : \beta < \alpha\}$ such that H_j contains a finite intersection $H_{\phi(\beta_1)} \cap \ldots \cap H_{\phi(\beta_k)}$, with $\beta_1, \ldots, \beta_k < \alpha$.

(II) If $\{j : H_j \geq H_{\phi(\beta_1)} \cap \ldots \cap H_{\phi(\beta_k)}$, for some $\beta_1, \ldots, \beta_k < \alpha\} \subseteq \{\phi(\beta) : \beta < \alpha\}$, then define $\phi(\alpha)$ to be the smallest $j \notin \{\phi(\beta) : \beta < \alpha\}$.

[That is, after introducing a new H_j we then include all H_k which contain a finite intersection of listed subgroups.]

For the purposes of this proof, we shall say that an ordinal $\alpha < \rho$ is *closed* if $\{j : H_j \geq H_{\phi(\beta_1)} \cap \ldots \cap H_{\phi(\beta_k)}$, for some $\beta_1, \ldots, \beta_k < \alpha\} \subseteq \{\phi(\beta) : \beta < \alpha\}$. If α is any infinite ordinal less than ρ, then the succeeding closed ordinal has the same cardinality as α. It is clear therefore that there are $|J|$ closed ordinals.

We show that there are $\exp|J|$ distinct cosets of H in L by forming $\exp|J|$ sequences of cosets $\{H_{\phi(\alpha)} x_\alpha : \alpha < \rho\}$ such that each finite subset $\{H_{\phi(\alpha_1)} x_{\alpha_1}, \ldots, H_{\phi(\alpha_k)} x_{\alpha_k}\}$ has a non-empty intersection, so that if x is an element of this intersection $H_{\phi(\alpha_i)} x = H_{\phi(\alpha_i)} x_{\alpha_i}$, for each $i = 1, \ldots, k$. Lemma 4.8 shows that, given such a sequence, the intersection of all the cosets $\bigcap_{\alpha<\rho} H_{\phi(\alpha)} x_\alpha$ is a coset of H in L and that distinct sequences give rise to distinct cosets of H.

A sequence $\{H_{\phi(\beta)} x_\beta : \beta < \alpha\}$ of cosets is called a *coherent* sequence of length α if each finite subset has non-empty intersection. We show that there are $\exp|J|$ distinct coherent sequences of length ρ by showing that each coherent sequence of length α can be extended to a coherent sequence of length $\alpha+1$ and that if α is a closed ordinal then each coherent sequence of length α can be extended in two different ways.

(A) Suppose that the coherent sequence $\{H_{\phi(\beta)} x_\beta : \beta < \alpha\}$ can not be extended.

Let the cosets of $H_{\phi(\alpha)}$ be $H_{\phi(\alpha)} y_1, \ldots, H_{\phi(\alpha)} y_k$; then, for each $m = 1, \ldots, k$, there is a finite set

$$\{H_{\phi(\beta_1(m))} x_{\beta_1(m)}, \ldots, H_{\phi(\beta_n(m))} x_{\beta_n(m)}\}$$

such that the set

$$\{H_{\phi(\alpha)}\, y_m,\ H_{\phi(\beta_1(m))}\, x_{\beta_1(m)},\ \ldots,\ H_{\phi(\beta_n(m))}\, x_{\beta_n(m)}\}$$

has empty intersection. But since $\{H_{\phi(\beta)}x_\beta : \beta < \alpha\}$ is coherent there is an element x in the intersection

$$\bigcap\{H_{\phi(\beta_r(s))}\, x_{\beta_r(s)} : 1 \leq r \leq n,\ 1 \leq s \leq k\} .$$

But the coset $H_{\phi(\alpha)}x$ must be one of the cosets $H_{\phi(\alpha)}y_1, \ldots, H_{\phi(\alpha)}y_k$ and so we have a contradiction.

(B) Suppose that α is closed and that the coherent sequence $\{H_{\phi(\beta)}x_\beta : \beta < \alpha\}$ can only be extended by adding the coset $H_{\phi(\alpha)}x_\alpha$.

Let the distinct cosets of $H_{\phi(\alpha)}$ be $H_{\phi(\alpha)}x_\alpha$, $H_{\phi(\alpha)}y_1, \ldots, H_{\phi(\alpha)}y_k$. Since the coherent sequence $\{H_{\phi(\beta)} : \beta < \alpha\}$ can not be extended by adding one of the cosets $H_{\phi(\alpha)}y_m$, there is, for each m = 1, ..., k, a finite set

$$\{H_{\phi(\beta_1(m))}\, x_{\beta_1(m)}, \ldots, H_{\phi(\beta_n(m))}\, x_{\beta_n(m)}\}$$

such that the set

$$\{H_{\phi(\alpha)}y_m,\ H_{\phi(\beta_1(m))}x_{\beta_1(m)}, \ldots, H_{\phi(\beta_n(m))}\, x_{\beta_n(m)}\}$$

has empty intersection. Since α is closed there is a $\gamma < \alpha$ such that $H_{\phi(\gamma)} \leq \bigcap\{H_{\phi(\beta_r(s))} : 1 \leq r \leq n,\ 1 \leq s \leq k\}$. Also, since α is closed, $H_{\phi(\alpha)} \not\geq H_{\phi(\gamma)}$ and so there is an element $g \in H_{\phi(\gamma)} - H_{\phi(\alpha)}$. Since $\{H_{\phi(\beta)} : \beta < \alpha\}$ is coherent there is an element x in the intersection

$$\bigcap\{H_{\phi(\beta_r(s)}\, x_{\beta_r(s)} : 1 \leq r \leq n,\ 1 \leq s \leq k\}.$$

Since $g \in H_{\phi(\gamma)}$, the element gx is also in this intersection. Since $g \notin H_{\phi(\alpha)}$ the cosets $H_{\phi(\alpha)}x$ and $H_{\phi(\alpha)}\, gx$ are distinct and so one of them is an $H_{\phi(\alpha)}y_m$. Thus one of the elements x and gx is in the intersection

$$H_{\phi(\alpha)}y_m \cap \bigcap\{ H_{\phi(\beta_r(m))}\, x_{\beta_r(m)} : 1 \leq r \leq n\}$$

and we have a contradiction.

This completes the construction of the coherent sequences and hence the proof of Theorem 4.9. □

In particular, this result gives a method of determining the cardinality of a profinite completion by taking the identity as the closed subgroup of L and using the correspondence between the subgroups N_i of G and K_i of L given in Lemma 4.6.

Corollary 4.10. Let G be a residually finite group with residual system $\mathcal{N} = \{N_i : i \in I\}$. If the N_i are distinct and L is the profinite completion of G with respect to $\mathcal{N}$, then $|L| = \exp |I|$. □

One of the consequences of this result is that a residually finite group may have profinite completions of different cardinalities. For example the cartesian product $G = \prod_{n=1}^{\infty} G_n$ of countably many groups each cyclic of order p is isomorphic to the direct product $Dr_{\lambda \in \Lambda} H_\lambda$ of $\exp \aleph_0$ groups each cyclic of order p. The group G is itself profinite with respect to the residual system $\{N_i : i < \omega\}$, where $N_i = \prod_{n=i+1}^{\infty} G_n$ and $|G| = \exp \aleph_0$. However, G also has a residual system $\{M_\sigma : \sigma \in \Sigma\}$, where Σ is the set of finite subsets of Λ and $M_\sigma = Dr_{\lambda \notin \sigma} H_\lambda$. Clearly $|\Sigma| = \exp \aleph_0$ and so the profinite completion of G with respect to $\{M_\sigma : \sigma \in \Sigma\}$ has cardinality $\exp \exp \aleph_0$.

We are of course particularly concerned with the profinite completion L of a residually finite periodic FC-group G. Some features of the embedding of G into L are rather nicer than in the general case.

Lemma 4.11. Let G be a residually finite periodic FC-group with a residual system $\mathcal{N} = \{N_i : i \in I\}$, let H be a subgroup of G and let L be the profinite completion of G with respect $\mathcal{N}$. Then

(i) each finite normal subgroup of G is normal in L,

(ii) $G \triangleleft L$,

(iii) $N_L(H)$ is a closed subgroup of L, and hence $N_G(H) = N_L(H) \cap G$ is a closed subgroup of G.

Proof. (i) If F is a finite normal subgroup of G, then F is a closed subgroup of L, by Lemma 4.2 (ii), and hence by Lemma 4.7, $N_L(F)$ is closed.

But $N_L(F) \geq G$ and G is dense in L, therefore $N_L(F) = L$ and $F \triangleleft L$.

(ii) The group G is generated by its finite normal subgroups each of which is normal in L and hence $G \triangleleft L$.

(iii) Let $\{F_i : i \in I\}$ be a local system of finite normal subgroups of G. For each $i \in I$, $F_i \cap H$ is finite and so, by Lemma 4.2, is a closed subgroup of L. By Lemma 4.7, $N_L(F_i \cap H)$ is a closed subgroup of L. But since each F_i is normal in L, it is clear that $N_L(H) = \bigcap_{i \in I} N_L(F_i \cap H)$, so that $N_L(H)$, being an intersection of closed subgroups, is itself closed. □

Our main application of the results outlined above is in the study of locally inner automorphisms of a (periodic) FC-group G. We shall see in the next two Chapters that many results from the theory of finite groups can be extended to (periodic) FC-groups and that, in these generalizations, inner automorphisms are replaced by locally inner automorphisms.

An automorphism ϕ of a group G is said to be *locally inner* if, for each finite set $\{g_1, \ldots, g_n\}$ of elements of G, there is an element $x \in G$ such that

$$g_i\phi = x^{-1} g_i x, \text{ for } i = 1,2,\ldots,n.$$

It is clear that the locally inner automorphisms of any group G form a group Linn G and that

$$\text{Inn } G \leq \text{Linn } G \leq \text{Aut } G.$$

If S is any subset of G and H a subgroup of G, then we can talk about the centralizer of S and normalizer of H in any subgroup A of Aut G. Thus $C_A(S) = \{\phi \in A : x\phi = x, \text{ for all } x \in S\}$ and $N_A(H) = \{\phi \in A : H\phi = H\}$. The group of inner automorphisms Inn G of any group G is isomorphic to G/Z(G) and under this isomorphism, $C_{\text{Inn } G}(S)$ corresponds to $C_G(S)/Z$ and $N_{\text{Inn } G}(H)$ corresponds to $N_G(H)/Z$. If G is an FC-group then the group of locally inner automorphisms is isomorphic to a profinite completion of G/Z(G).

Theorem 4.12. Let $\{F_i : i \in I\}$ be a local system of finitely generated normal subgroups of the FC-group G and, for each $i \in I$, let $C_i = C_G(F_i)$. Then $\{C_i/Z : i \in I\}$ is a residual system of G/Z and Linn G is isomorphic to the profinite completion of G/Z with respect to $\{C_i/Z : i \in I\}$.

If L = Linn G and $\pi_i : L \to G/C_i$ is the projection map, then Ker $\pi_i = C_L(F_i)$.

Proof. We proved in Lemma 1.11 that $\{C_i/Z : i \in I\}$ is a residual system of G/Z.

Let $A = G/Z \cong$ Inn G and L = Linn G; then there is a natural embedding of A in L in which the element gZ corresponds to the inner automorphism of G induced by conjugation by g. For each $i \in I$, each element of L coincides with an inner automorphism of G on F_i. Therefore, if $D_i = C_L(F_i)$, we have $L = AD_i$. Therefore $L/D_i \cong A/C_A(F_i) \cong G/C_i$ and we have a natural homomorphism π_i from L onto G/C_i with Ker $\pi_i = D_i$. The homomorphisms π_i induce a homomorphism $\theta : L \to \Pi_{i \in I}(G/C_i)$. If the locally inner automorphism ϕ coincides with the inner automorphism induced by g_i on F_i then $(\phi)\theta = (g_i C_i)$. It is clear that Im $\theta \leq \varprojlim (G/C_i)$.

Let ϕ be a non-trivial locally inner automorphism; then there is an $x \in G$ such that $x\phi \neq x$. Clearly $(\phi)\theta \neq 1$ and so Ker $\theta = 1$.

Let $(g_i C_i) \in \varprojlim (G/C_i)$. Define $x\phi = g_i^{-1} x g_i$, where $x \in F_i$. This definition is consistent, for if $x \in F_i \cap F_j$ then there is a $k \in I$ such that $F_i F_j \leq F_k$ and hence $C_k \leq C_i \cap C_j$. Therefore $g_k C_i = g_i C_i$ and $g_k C_j = g_j C_j$; hence $g_i^{-1} x g_i = g_k^{-1} x g_k = g_j^{-1} x g_j$. It is clear that $\phi \in$ Linn G and that $(\phi)\theta = (g_i C_i)$. Thus Im $\theta = \varprojlim (G/C_i)$ and so Linn $G \cong \varprojlim (G/C_i)$.□

If G is a residually finite periodic FC-group then Linn G can be constructed by giving a residual system of G/Z derived from a given residual system of G.

Theorem 4.13. Let G be a residually finite periodic FC-group and let $\{N_i : i \in I\}$ be a residual system of G. If $Z_i/N_i = Z(G/N_i)$, then Linn G is isomorphic to $\varprojlim (G/Z_i)$.

If $\pi_i : L \to G/Z_i$ is the projection map, then Ker $\pi_i = C_L(G/N_i)$.

Proof. We have already seen in Lemma 1.11 that $\{Z_i/Z : i \in I\}$ is a residual system of G/Z. We show that this system is equivalent to the residual system $\{C_G(F_j)/Z : j \in J\}$ used in the previous theorem, where $\{F_j : j \in J\}$ is the local system consisting of all finite normal subgroups of G.

If F is a finite normal subgroup of G then there is a subgroup N_i such

that $N_i \cap F = 1$. Then $[Z_i, F] \leq N_i \cap F = 1$ and so $Z_i \leq C_G(F)$. Conversely, given $i \in I$, there is a finite normal subgroup F of G such that $FN_i = G$. Thus $[C_G(F), G] = [C_G(F), FN_i] = [C_G(F), N_i] \leq N_i$ and so $C_G(F) \leq Z_i$. It follows from Lemma 4.5(i) that the two residual systems $\{Z_i/Z : i \in I\}$ and $\{C_G(F_j)/Z : j \in J\}$ are equivalent and so, using Lemma 4.5(ii) and Theorem 4.12,
$\text{Linn } G \cong \varprojlim(G/C_G(F_j)) \cong \varprojlim (G/Z_i)$.

If F is a finite normal subgroup of G such that $FN_i = G$ then $G/N_i \cong F/(F \cap N_i)$ and $Z_i = C_G(F/(F \cap N_i))$. The mapping $\pi_i : L \to G/Z_i$ may be thought of as the composition of the natural mappings

$$L \to G/C_G(F) \to G/Z_i$$

from which we can see that, for $\phi \in L$, $\phi\pi_i$ is the automorphism of $F/(F \cap N_i)$ or G/N_i induced by ϕ and so $\text{Ker } \pi_i = C_L(G/N_i)$. □

The characterization of the group of locally inner automorphisms of an FC-group given in Theorem 4.12 together with the results on the cardinality of a profinite group can be combined to give the following theorem.

Theorem 4.14. (Robinson, Stonehewer and Wiegold [85]). *Let G be an FC-group.*

(i) *If* $G/Z(G)$ *is finite, then* $\text{Linn } G = \text{Inn } G \cong G/Z(G)$

(ii) *If* $G/Z(G)$ *is infinite, then* $|\text{Linn } G| = \exp |\text{Inn } G|$.

Proof. (i) is obvious.

(ii) Let $\{F_i : i \in I\}$ be a local system of finitely generated normal subgroups of G and, for each $i \in I$, let $C_i = C_G(F_i)$. By Theorem 4.12 and Corollary 4.10, it is sufficient to prove that the set of distinct C_i's has cardinality $|\text{Inn } G|$.

Let $F_i^* = C_G(C_i)$; then $C_G(F_i^*) = C_i$ and so $C_i = C_j$ if and only if $F_i^* = F_j^*$. Thus the cardinality of the set of distinct C_i's is equal to the cardinality of the set of distinct F_i^*'s. Clearly $\{F_i^*/Z : i \in I\}$ is a local system of normal subgroups of G/Z. Also there is a finitely generated normal subgroup X of G such that $XC_i = G$. We have $F_i^* \cap C_G(X) = C_G(C_i) \cap C_G(X) = Z$; but $G/C_G(X)$ is finite and so F_i^*/Z must be finite. Thus $\{F_i^*/Z : i \in I\}$ is a local system of finite normal subgroups of G/Z and there must be |G/Z| distinct subgroups F_i^*. The result now follows from the isomorphism between G/Z and Inn G.□

Corollary 4.15. (Stonehewer [95]) Let G be a p-group which is an FC-group. Then G has an outer automorphism (unless G has order p or is the trivial group).

Proof. If G/Z(G) is infinite, then |Linn G| > |Inn G|, by Theorem 4.14.

If G/Z(G) is finite, then there is a finite normal subgroup F of G such that G = FZ and |F| > p. By Gaschütz's Theorem ([55], p.403), F has an outer automorphism θ which fixes each element of Z(F) and hence of $F \cap Z$. Define an automorphism ϕ of G by

$$(xz)\phi = (x\theta)z \ , \text{ for all } x \in F, \ z \in Z.$$

Since θ fixes each element of $F \cap Z$, this is well-defined and is clearly an outer automorphism of G. □

Stonehewer actually proved that G has an outer automorphism of order p. This sharper result may also be obtained by the above methods making a simple variation in the proof of Theorem 4.9 so that the only cosets of the subgroups K_i which are considered are those of order p. As the group of locally inner automorphisms of an FC-group seems to correspond to the group of inner automorphisms of a finite group, the following question is of interest.

Question 4A. If G is a p-group which is an FC-group, does G have an automorphism (of order p) which is not locally inner?

Our results in the next two Chapters will depend on building up results in periodic FC-groups from known results for finite soluble groups. Many of these results are conjugacy theorems in finite groups (e.g. Sylow's Theorem) and we need to build up locally inner automorphisms from inner automorphisms of finite groups, using either the finite normal subgroups or the finite factor groups.

Theorem 4.16. (Stonehewer [93]) Let $\{F_i : i \in I\}$ be a local system of finitely generated normal subgroups of the FC-group G. For each $i \in I$, let A_i be a non-empty set of automorphisms of F_i induced by inner automorphisms of G such that, whenever $F_i \geq F_j$, each element of A_i induces in F_j an automorphism which is in A_j. Then there is a locally inner automorphism of G inducing in each F_i an automorphism from A_i.

Proof. Let $C_i = C_L(F_i)$, where L = Linn G; then $C_i = \text{Ker}\,(\pi_i : L \to G/C_G(F_i))$ is closed in L. Let $B_i = \{\phi \in L : \phi|_{F_i} \in A_i\}$; then B_i is a finite union of cosets of C_i in L and so is a closed subset of L. If $F_i \geq F_j$, then $B_i \subseteq B_j$.

Given a finite collection $B_{i_1}, \ldots, B_{i_k}$, there is an $r \in I$ such that $F_{i_1} \ldots F_{i_k} \leq F_r$ and so $B_r \subseteq B_{i_1} \cap \ldots \cap B_{i_k}$. Thus the family $\{B_i : i \in I\}$ has the finite intersection property and, since L is compact, $\bigcap_{i \in I} B_i \neq \emptyset$. Thus there is a locally inner automorphism $\phi \in \bigcap_{i \in I} B_i$ and so ϕ induces an automorphism from A_i in each B_i. □

Theorem 4.17. (Tomkinson [99]) Let $\{N_i : i \in I\}$ be a residual system of normal subgroups of finite index in the residually finite periodic FC-group G. For each $i \in I$, let A_i be a non-empty set of inner automorphisms of G/N_i such that, whenever $N_i \leq N_j$, each element of A_i induces in G/N_j an automorphism which is in A_j. Then there is a locally inner automorphism of G inducing in each G/N_i an automorphism from A_i.

Proof. Let $Z_i = C_G(G/N_i)$, so that Linn G $\cong \varprojlim (G/Z_i)$, by Theorem 4.13, and if $\pi_i : L \to G/Z_i$ is the projection map then Ker $\pi_i = C_L(G/N_i)$ is a closed subgroup of L. Let $B_i = \{\phi \in L$: the automorphism induced in G/N_i by ϕ is in $A_i\}$; then B_i is the union of finitely many cosets of $C_L(G/N_i)$ in L and so is a closed subset of L. Also if $N_i \leq N_j$, then $B_i \subseteq B_j$.

Given a finite collection $B_{i_1}, \ldots, B_{i_k}$, there is an $r \in I$ such that $N_r \leq N_{i_1} \cap \ldots \cap N_{i_k}$ and so $B_r \subseteq B_{i_1} \cap \ldots \cap B_{i_k}$. Thus the family $\{B_i : i \in I\}$ has the finite intersection property and, since L is compact, $\bigcap_{i \in I} B_i \neq \emptyset$. Thus there is a locally inner automorphism $\phi \in \bigcap_{i \in I} B_i$ and so ϕ induces an automorphism from A_i in each G/N_i. □

It is obvious that an inner automorphism can be extended from a subgroup

H to a group G or lifted from a factor group G/N to a group G. This is also true of locally inner automorphisms in an FC-group G.

Theorem 4.18. (Stonehewer [93], Tomkinson [98]) *Let G be an FC-group.*

(i) *If* $H \leq G$ *and* $\phi \in \text{Linn } H$, *then there is a* $\theta \in \text{Linn } G$ *such that* $\theta|_H = \phi$.

(ii) *If* $N \triangleleft G$ *and* $\phi \in \text{Linn }(G/N)$, *then there is a* $\theta \in \text{Linn } G$ *such that* θ *induces* ϕ *in* G/N.

Proof. Let $\{F_i : i \in I\}$ be a local system of finitely generated normal subgroups of G.

(i) Let A_i be the set of automorphisms of F_i induced by inner automorphisms of G which coincide with ϕ on $H \cap F_i$. Then the sets A_i satisfy the conditions of Theorem 4.16 and so there is a locally inner automorphism θ of G coinciding with ϕ on each $H \cap F_i$ and hence on the whole of H.

(ii) Let A_i be the set of automorphisms of F_i induced by inner automorphisms of G such that the automorphism induced in F_iN/N coincides with ϕ. Again the sets A_i satisfy the conditions of Theorem 4.16 and so there is a locally inner automorphism θ of G such that the automorphism induced by θ in G/N coincides with ϕ on each F_iN/N and hence on the whole of G/N. □

Two subgroups H,K of a group G are said to be *locally conjugate* in G if there is a locally inner automorphism ϕ of G such that $H\phi = K$. We note first that in an FC-group G a subgroup can not be locally conjugate to a proper subgroup of itself. In general, of course, a subgroup of an infinite group may even be conjugate to a proper subgroup of itself.

Lemma 4.19. *Let H be a subgroup of the FC-group G and* $\phi \in \text{Linn } G$. *If* $H\phi \leq H$, *then* $H\phi = H$.

Proof. Let $\{F_i : i \in I\}$ be a local system of finitely generated normal subgroups of G. For each $i \in I$, there is an element $x \in G$ such that $x_i^{-1}(H \cap F_i)x_i = (H \cap F_i)\phi = H\phi \cap F_i \leq H \cap F_i$. Let Z be the centre of G; then $F_i/(F_i \cap Z)$ is finite and $H \cap F_i \cap Z \triangleleft G$. Thus $x_i^{-1}(H \cap F_i)x_i/(H \cap F_i \cap Z) \leq (H \cap F_i)/(H \cap F_i \cap Z)$. But $(H \cap F_i)/(H \cap F_i \cap Z) \cong (H \cap F_i)(F_i \cap Z)/(F_i \cap Z)$ is finite and so $x_i^{-1}(H \cap F_i)x_i = H \cap F_i$. Hence $H\phi = \bigcup_{i \in I}(H \cap F_i)\phi = \bigcup_{i \in I} x_i^{-1}(H \cap F_i)x_i = \bigcup_{i \in I}(H \cap F_i) = H$. □

If H is a subgroup of the group G, then we use $C\ell(H)$ (or $C\ell_G(H)$) to denote the *conjugacy class* containing H, that is the set of all subgroups of G which are conjugate to H in G. The set of all subgroups locally conjugate to H in G, the *local conjugacy class* containing H, is denoted by $Lc\ell(H)$ (or $Lc\ell_G(H)$).

Most conjugacy theorems for finite groups lead to a corresponding local conjugacy theorem for (periodic) FC-groups and there is some interest in determining when locally conjugate subgroups are actually conjugate or when a local conjugacy class is a conjugacy class.

Lemma 4.20. *Let H and K be conjugate subgroups of the FC-group G. Then $|H:H\cap K|$ is finite.*

Proof. Suppose $K = x^{-1}Hx$. Then $H\cap C_G(x)$ is invariant under x and so $H\cap C_G(x) \leqslant H\cap K$. But $|G:C_G(x)|$ is finite and hence $|H:H\cap K|$ is finite. □

Lemma 4.21. *Let H be a subgroup of the FC-group G.*

(i) *$C\ell_G(H)$ is finite if and only if H/H_G is finite.*

(ii) *If $C\ell_G(H)$ is finite, then $Lc\ell_G(H) = C\ell_G(H)$.*

Proof. (i) Suppose $C\ell_G(H)$ is finite; then it follows from Lemma 4.20 that H/H_G is finite. Conversely, suppose H/H_G is finite; then the finite subgroup H/H_G can have only finitely many conjugates in G/H_G. Hence $C\ell_G(H)$ is finite.

(ii) By part (i), we have H/H_G is finite. Thus $H = \langle H_G, h_1, \ldots, h_n\rangle$ and, for any $\phi \in \text{Linn } G$, $H\phi = \langle H_G, h_1\phi, \ldots, h_n\phi\rangle$. There is an element $x \in G$ such that $h_i\phi = x^{-1}h_ix$, for each $i = 1,2,\ldots,n$, and so $H\phi = \langle H_G, x^{-1}h_1x, \ldots, x^{-1}h_nx\rangle = x^{-1}Hx$. Therefore each subgroup locally conjugate to H is actually conjugate to H. □

We consider the question of when there are converses to these two lemmas. More precisely, if H and K are locally conjugate subgroups of G with $|H:H\cap K|$ finite are H and K conjugate in G? Secondly, if $Lc\ell(H) = C\ell(H)$ does it follow that $C\ell(H)$ is finite?

For the first of these questions we introduce a condition which we will see in later chapters applies to many of the important subgroups which we introduce. A subgroup H of G is said to be L-*pronormal* if, for each $\phi \in \text{Linn } G$, H and $H\phi$ are locally conjugate in their join $\langle H, H\phi \rangle$. One could define pronormality relative to any set of automorphisms, the usual concept of pronormality is then pronormality relative to the group of inner automorphisms; that is, H is *pronormal* in G if, for each $x \in G$, H and $x^{-1}Hx$ are conjugate in their join $\langle H, x^{-1}Hx \rangle$.

Theorem 4.22. (Tomkinson [100]) Let H be an L-pronormal subgroup of the FC-group G and let ϕ be a locally inner automorphism of G. Then H and $H\phi$ are conjugate in G if and only if $|H : H \cap H\phi|$ is finite.

Proof. One direction is given in Lemma 4.20. So assume that $|H:H \cap H\phi|$ is finite. There is a finitely generated normal subgroup F of G such that $F(H\phi) \geq FH$. But $F(H\phi) = (FH)\phi$ is locally conjugate to FH and so $F(H\phi) = FH$, by Lemma 4.19. Since H is L-pronormal in G, H and $H\phi$ are locally conjugate in $\langle H, H\phi \rangle$ and hence, by Theorem 4.18, in FH. Let Z be the centre of FH; then FZ/Z is finite. Also $N_{FH}(H) \geq HZ$ and so

$$|C\ell_{FH}(H)| \leq |FH:HZ| = |F:F \cap HZ| \leq |F:F \cap Z| = |FZ/Z| .$$

Hence $C\ell_{FH}(H)$ is finite and so, by Lemma 4.21, H and $H\phi$ are conjugate in FH and hence in G. □

The condition of L-pronormality is satisfied by Sylow subgroups, $\mathfrak{F}$-projectors and $\mathfrak{X}$-injectors and so the above result applies to all these important classes of subgroups. It is however fairly easy to construct examples in which $|H:H \cap H\phi|$ is finite but H and $H\phi$ are not conjugate. More complicated examples with further restrictions on H have been given by Hartley and Parker [53].

Example 4.23. (Tomkinson [98]) Let X be the direct product of countably many groups, each isomorphic to the four-group: $X = \text{Dr}_{n=0}^{\infty} V_n$, where $V_n = \langle a_n, b_n : a_n^2 = b_n^2 = (a_n b_n)^2 = 1 \rangle$.

For each $n > 0$, X has an automorphism α_n of order 2 defined by

$$a_m\alpha_n = \begin{cases} b_m, & \text{if } m = 0 \text{ or } n, \\ a_n, & \text{if } m \neq 0 \text{ or } n, \end{cases} \qquad b_m\alpha_n = \begin{cases} a_m, & \text{if } m = 0 \text{ or } n, \\ b_m, & \text{if } m \neq 0 \text{ or } n. \end{cases}$$

Let $Y = \mathrm{Dr}_{n=1}^{\infty} \langle \alpha_n \rangle \leq \mathrm{Aut}\, X$ and form the split extension G of X by Y. Then G is generated by the finite normal subgroups $F_n = \langle V_o, \ldots, V_n, \alpha_1, \ldots, \alpha_n \rangle$ and so is a countable periodic FC-group.

Let $H = \mathrm{Dr}_{n=0}^{\infty} \langle a_n \rangle$ and $K = \langle b_o \rangle \times \mathrm{Dr}_{n=1}^{\infty} \langle a_n \rangle$; then $K \cap F_n = (H \cap F_n)\alpha_{n+1}$ so that H and K are locally conjugate in G and $|H:H \cap K| = 2$. But if x is any element of G, then $|H:H \cap x^{-1}Hx| \geq 4$; so it is clear that H and K are not conjugate in G. □

One other important class of subgroups to be introduced in the next Chapter is that of the basis normalizers. These are not, in general, L-pronormal but we have been unable to determine whether the above example or one of those given in [53] can be adapted so that H is a basis normalizer. The example which we claimed to give in [100] is incorrect.

Question 4B. Is there a periodic FC-group G with basis normalizers H and K such that $|H:H \cap K|$ is finite but H and K are not conjugate in G?

The relationship between the two conditions pronormality and L-pronormality is somewhat unclear. The condition of L-pronormality appears to be stronger than pronormality but we have been unable to construct an example to show that this is the case.

Lemma 4.24. *Let H be an L-pronormal subgroup of the FC-group G. Then H is a pronormal subgroup of G.*

Proof. The conjugate subgroups H and $x^{-1}Hx$ are locally conjugate in their join $\langle H, x^{-1}Hx \rangle$, since H is L-pronormal. But $|H:H \cap x^{-1}Hx|$ is finite, by Lemma 4.20, and hence, by Theorem 4.22, H and $x^{-1}Hx$ are conjugate in their join. □

Question 4C. Is there a (periodic) FC-group G with a pronormal subgroup H such that H is not L-pronormal in G?

The question of when the converse to Lemma 4.21 is true has a rather more satisfactory answer. This stems from the following rather remarkable result which says that the local conjugacy class containing a subgroup H is always as large as it possibly can be.

Theorem 4.25. (Hartley [51]) Let H be a subgroup of the FC-group G such that H/H_G is finite. Then $|Lc\ell(H)| = \exp(|H/H_G|)$.

Proof. It is clear that $|Lc\ell_G(H)| = |Lc\ell_{G/H_G}(H/H_G)|$ and so we may factor out H_G and so assume that $H_G = 1$; in particular, this implies that $H \cap Z = 1$ and so $H \cong HZ/Z$ is a periodic group. It is also clear that $|Lc\ell(H)| \leqslant \exp(|H|)$ and so we must show that $|Lc\ell(H)| \geqslant \exp(|H|)$.

Let $\{F_i : i \in I\}$ be a local system of finitely generated normal subgroups of G and, for each $i \in I$, let $C_i = C_G(F_i)$. Then L = Linn G is the profinite completion of G/Z with respect to the residual system $\{C_i/Z : i \in I\}$. Theorem 4.12 also shows that the standard residual system of L consists of the subgroups $K_i = C_L(F_i)$, $i \in I$.

Now $|Lc\ell(H)| = |L:N_L(H)|$ and since, by Lemma 4.11, $N_L(H)$ is a closed subgroup of L, Theorem 4.9 shows that the index $|L:N_L(H)|$ is equal to $\exp \mathfrak{m}$, where $\mathfrak{m}$ is the cardinality of the set of distinct subgroups $K_iN_L(H)$. For each $i \in I$, $K_iN_L(H) \leqslant N_L(F_i \cap H)$ and since there are only finitely many subgroups of L containing each $K_iN_L(H)$, it is sufficient to show that there are $|H|$ distinct subgroups $N_L(F_i \cap H)$. This will be the case if there are $|H|$ distinct subgroups $N_G(F_i \cap H) = G \cap N_L(F_i \cap H)$.

For each $i \in I$, let $H_i = \langle F_j \cap H : N_G(F_j \cap H) = N_G(F_i \cap H) \rangle$. Then $N_G(H_i) \leqslant N_G(F_i \cap H_i) = N_G(F_i \cap H)$ and, conversely, $N_G(F_i \cap H)$ normalizes all the $F_j \cap H$ which generate H_i. Therefore $N_G(H_i) = N_G(F_i \cap H)$. Hence $N_G(F_i \cap H) = N_G(F_j \cap H)$ if and only if $H_i = H_j$. Also $|G:N_G(H_i)|$ is finite and so $H_i/(H_i)_G$ is finite, using Lemma 4.21(i). Since $H_G = 1$, $(H_i)_G$ is also trivial and so H_i is finite. But clearly H is generated by the H_i and so there are $|H|$ distinct subgroups H_i and hence $|H|$ distinct subgroups $N_G(F_i \cap H)$. □

Precise information can also be given about the size of a conjugacy class for groups in the class $\mathfrak{Z}$ introduced in Chapter 3. The converse to Lemma 4.21 can be proved for the class $\mathfrak{Y}$ which we recall is possibly larger than the class of periodic $\mathfrak{Z}$-groups.

Theorem 4.26. (i) Let $G \in \mathfrak{Z}$ and $H \leq G$. If H/H_G is infinite then $|C\ell(H)| = |H/H_G|$.

(ii) Let $G \in \mathfrak{Y}$ and $H \leq G$. Then $Lc\ell(H) = C\ell(H)$ if and only if $C\ell(H)$ is finite.

Proof. (i) As in the previous theorem, we may assume that $H_G = 1$ and H is infinite. Since $G \in \mathfrak{Z}$, $|G:C_G(H)| \leq |H|$ and so $|C\ell(H)| = |G:N_G(H)| \leq |H|$. By Lemma 1.19, there is a normal subgroup N of G such that $NC_G(H) = G$ and $|N| \leq |C\ell(H)|$. Now $C_H(N) \lhd NC_G(H) = G$ and so $C_H(N) = 1$. Therefore $|H| = |H/C_H(N)| \leq |G:C_G(N)| \leq |N| \leq |C\ell(H)|$, giving the required equality.

(ii) We know from Lemma 4.21 that if $C\ell(H)$ is finite, then $Lc\ell(H) = C\ell(H)$. So suppose that $C\ell(H)$ is infinite and choose elements $x_1, x_2, \ldots$ in G so that the conjugates $x_1^{-1}H x_1, x_2^{-1}Hx_2, \ldots$ are distinct. For each pair (i,j) choose an element $h_{ij} \in H$ such that $x_i^{-1} h_{ij} x_i \notin x_j^{-1} Hx_j$. Let K be the normal subgroup of G generated by the x_i and the h_{ij}; then K is countably infinite and $H \cap K$ has infinitely many conjugates in K. It follows from Lemma 4.21(i) that $(H \cap K)/(H \cap K)_G$ is infinite and hence $Lc\ell_G(H \cap K)$ is uncountable, by Theorem 4.25. But since $G \in \mathfrak{Y}$ and $H \cap K$ is countable, $C\ell_G(H \cap K)$ is countable. Thus $Lc\ell(H \cap K) \neq C\ell(H \cap K)$ and hence $Lc\ell(H) \neq C\ell(H)$. □

We stress again that we know of no $\mathfrak{Y}$-group which is not a $\mathfrak{Z}$-group. One would therefore expect that it might be possible to extend part (i) of the above theorem to $\mathfrak{Y}$-groups.

Question 4D. If H is a subgroup of the $\mathfrak{Y}$-group G such that H/H_G is infinite, is $|C\ell(H)| = |H/H_G|$?

Example 3.8 shows that part (ii) of Theorem 4.26 is probably as far as one can go in giving a converse to Lemma 4.21. In that group, each subgroup locally conjugate to X has the form $X\phi = \mathrm{Dr}_{n=1}^{\infty} \langle x_n^* \rangle$, where x_n^* is one of the elements $x_n, x_n z, \ldots, x_n z^{p-1}$. Clearly there is an element $y \in Y$ such that $X\phi = y^{-1}Xy$ and so $Lc\ell(X) = C\ell(X)$ even though $C\ell(X)$ is infinite.

However, the results given in Theorem 4.26 can be applied to most of the interesting (local) conjugacy classes which arise in the following Chapters since a residually finite periodic FC-group is a $\mathfrak{Z}$-group and so this theorem can always be applied in the factor group $G/Z(G)$.

Perhaps the most surprising feature of Theorem 4.25 is that the size of the local conjugacy class $Lc\ell(H)$ is determined entirely by $|H/H_G|$ and is not directly connected with the size of the conjugacy class $C\ell(H)$. We can of course deduce certain inequalities.

Theorem 4.27. (Hartley [51]) Let H be a subgroup of the FC-group G such that H/H_G is infinite. Then

(i) $|C\ell(H)| \leqslant \exp(|H/H_G|)$

(ii) $|H/H_G| \leqslant \exp |C\ell(H)|$, and hence $|Lc\ell(H)| \leqslant \exp \exp |C\ell(H)|$.

Proof. (i) By Theorem 4.25, $\exp(|H/H_G|) = |Lc\ell(H)| \geqslant |C\ell(H)|$.

(ii) For each $x \in G$, $|H:H \cap x^{-1}Hx|$ is finite, by Lemma 4.20, and so $H \cap x^{-1}Hx$ contains a normal subgroup H_x of H such that $H_G \leqslant H_x \leqslant H$ and H/H_x is finite. Since H_G is the intersection of all conjugates of H, $H_G = \bigcap_{x \in G} H_x$ and so H/H_G can be embedded in the cartesian product of the finite groups H/H_x. There are at most $|C\ell(H)|$ distinct subgroups H_x and so $|H/H_G| \leqslant$ $\leqslant \exp |C\ell(H)|$. □

Example 3.8 shows that both of these bounds can in fact be attained. For $|X/X_G| = \aleph_0$ but $|C\ell(X)| = \exp \aleph_0 = |Lc\ell(X)|$. On the other hand $|Y/Y_G| = \exp \aleph_0$ but $|C\ell(Y)| = |G:Y| = \aleph_0$ and Theorem 4.25 shows that $|Lc\ell(Y)| = \exp \exp \aleph_0$.

5 Sylow theory

In this Chapter we shall see how theorems about periodic FC-groups can be built up from known theorems about finite groups. The results given here are obtained most easily by considering a local system of finite normal subgroups. In the following Chapter the results are obtained by considering the finite factor groups corresponding to a residual system of normal subgroups of finite index in a residually finite periodic FC-group and so we shall also need to relate the results obtained in this chapter to residual systems as well as local systems.

As is usual when discussing infinite groups, a p-*group* is defined to be a group in which each element has order a power of p (where p is some prime). If π is a set of primes, then a π-*group* is a group in which each element has finite order and that order is a product of primes from the set π .

We define a *Sylow* p-*subgroup* (*Sylow* π-*subgroup*) of a group G to be a maximal p-subgroup (π-subgroup) of G; such subgroups necessarily exist by Zorn's Lemma. The main part of Sylow's Theorem then says that the Sylow p-subgroups of a finite group G are conjugate in G. Hall's Theorem says that, for any set of primes π, the Sylow π-subgroups of a finite soluble group G are conjugate in G.

To avoid stating results separately for Sylow p-subgroups and Sylow π-subgroups we shall say that a periodic FC-group G is a $\mathfrak{D}_\pi$-group if, in each finite subgroup F of G, the Sylow π-subgroups of F are conjugate in F. Thus a periodic FC-group is a $\mathfrak{D}_p$-group, for every prime p, and a locally soluble periodic FC-group is a $\mathfrak{D}_\pi$-group, for every set of primes π. If the Sylow π-subgroups of each subgroup of a finite group F are conjugate, then it is easy to prove that the Sylow π-subgroups of any factor group F/N are conjugate. Hence any section of a $\mathfrak{D}_\pi$-group is also a $\mathfrak{D}_\pi$-group.

Lemma 5.1. *Let S be a Sylow π-subgroup of the $\mathfrak{D}_\pi$-group G. If F is a finite normal subgroup of G, then S ∩ F is a Sylow π-subgroup of F.*

Proof. Clearly S is a Sylow π-subgroup of FS and |FS:S| is finite. Thus there is a normal subgroup N of FS such that $N \leq S$ and FS/N is finite. It follows that S/N is a Sylow π-subgroup of the finite $\mathfrak{D}_\pi$-group FS/N and hence $(S \cap FN)/N$ is a Sylow π-subgroup of the normal subgroup FN/N. (It is this step that requires the conjugacy of the Sylow π-subgroups of FS/N.) From the isomorphism $FN/N \cong F/(F \cap N)$, we obtain that $(S \cap F)/(F \cap N)$ is a Sylow π-subgroup of $F/(F \cap N)$ and hence $S \cap F$ is a Sylow π-subgroup of F. □

Theorem 5.2. (Baer [3], Gol'berg [40]) The Sylow π-subgroups of a $\mathfrak{D}_\pi$-group G are locally conjugate in G.

Proof. Let S and T be Sylow π-subgroups of G and let $\{F_i : i \in I\}$ be a local system of finite normal subgroups of G. Then, by Lemma 5.1, $S \cap F_i$ and $T \cap F_i$ are Sylow π-subgroups of F_i and so are conjugate. Now let A_i be the set of automorphisms of F_i induced by inner automorphisms of G which map $S \cap F_i$ onto $T \cap F_i$. Clearly each A_i is non-empty and if $F_i \geq F_j$, then each element of A_i induces an automorphism in F_j which is in the set A_j. Therefore we can apply Theorem 4.16 to obtain a locally inner automorphism ϕ of G such that $(S \cap F_i)\phi = T \cap F_i$, for each $i \in I$, and hence $S\phi = \bigcup_{i \in I}(S \cap F_i)\phi = \bigcup_{i \in I}(T \cap F_i) = T$. □

This proof consisted of taking a given Sylow π-subgroup S of G and considering its finite subgroups $S \cap F_i$ which are Sylow π-subgroups of the finite groups $F_i, i \in I$. Some of our results will depend on the reverse procedure, constructing a Sylow π-subgroup of G from given Sylow π-subgroups of finite normal subgroups of G.

Lemma 5.3. Let $\{F_i : i \in I\}$ be a local system of finite normal subgroups of the periodic FC-group G and, for each $i \in I$, let S_i be a Sylow π-subgroup of F_i such that, whenever $F_i \geq F_j$, $S_i \cap F_j$ is equal to S_j. Then $S = \bigcup_{i \in I} S_i$ is a Sylow π-subgroup of G.

Proof. It is clear that S is a π-subgroup of G and that $S \cap F_i = S_i$, for each $i \in I$. If T is a Sylow π-subgroup of G containing S, then $T \cap F_i$ is a π-subgroup of F_i containing S_i and so $T \cap F_i = S_i$ and $S = \bigcup_{i \in I} S_i = \bigcup_{i \in I}(T \cap F_i) = T$. □

Having proved the local conjugacy of Sylow π-subgroups we can now improve on Lemma 5.1 and show that Sylow π-subgroups behave well with respect to all normal subgroups and factor groups.

Theorem 5.4. *Let S be a Sylow π-subgroup and N a normal subgroup of the $\mathfrak{D}_\pi$-group G. Then*

(i) $S \cap N$ is a Sylow π-subgroup of N,

(ii) SN/N is a Sylow π-subgroup of G/N,

(iii) each Sylow π-subgroup of G/N is of the form TN/N, where T is a Sylow π-subgroup of G.

Proof. (i) Let W be a Sylow π-subgroup of N containing $S \cap N$; then W is contained in a Sylow π-subgroup T of G and $W = T \cap N$. By Theorem 5.2, there is a locally inner automorphism ϕ of G such that $T = S\phi$ and hence $W = T \cap N = S\phi \cap N = (S \cap N)\phi$. It now follows from Lemma 4.19 that $W = S \cap N$ and so $S \cap N$ is a Sylow π-subgroup of N.

(ii) Let $\{F_i : i \in I\}$ be a local system of finite normal subgroups of G; then $\{F_iN/N : i \in I\}$ forms a local system of finite normal subgroups of G/N. By Lemma 5.1, $S \cap F_i$ is a Sylow π-subgroup of F_i and hence $(S \cap F_i)(N \cap F_i)/(N \cap F_i)$ is a Sylow π-subgroup of $F_i/(N \cap F_i)$. Under the isomorphism between $F_i/(N \cap F_i)$ and F_iN/N, we see that $(S \cap F_i)N/N$ is a Sylow π-subgroup of F_iN/N. Writing $T_i = (S \cap F_i)N$, we have Sylow π-subgroups T_i/N of F_iN/N. Whenever $F_iN \geq F_jN$, $(T_i/N) \cap (F_jN/N)$ is a π-subgroup of F_jN/N containing the Sylow π-subgroup T_j/N and hence $(T_i/N) \cap (F_jN/N) = T_j/N$. It follows from Lemma 5.3 that $T/N = \bigcup_{i\in I}(T_i/N)$ is a Sylow π-subgroup of G/N and $T = \bigcup_{i\in I}(S \cap F_i)N = SN$, as required.

(iii) Let U/N be a Sylow π-subgroup of G/N and let S be a Sylow π-subgroup of G. Then, by (ii), SN/N is a Sylow π-subgroup of G/N and so there is a locally inner automorphism ϕ of G/N such that $U/N = (SN/N)\phi$. By Theorem 4.18(ii), there is a locally inner automorphism θ of G inducing ϕ in G/N. Thus $U = (SN)\theta = (S\theta)N$ so that $U/N = (S\theta)N/N$, where $S\theta$ is a Sylow π-subgroup of G. □

We could have proved the local conjugacy of Sylow π-subgroups in a residually finite $\mathfrak{D}_\pi$-group using a dual approach. Corresponding to Lemma 5.1, we could first have shown that SN/N is a Sylow π-subgroup of G/N for each finite factor group G/N and then proved Theorem 5.2 by using Theorem 4.17. This dual approach would have the disadvantage that to obtain the general result we would first have to consider G/Z(G) and then to include the centre. Although there are these disadvantages it will be necessary to consider residual systems in the next Chapter and so we obtain a result corresponding to Lemma 5.3.

Lemma 5.5. Let $N_i, i \in I$, be any normal subgroups of the $\mathfrak{D}_\pi$-group G and let S be a Sylow π-subgroup of G. Then $\bigcap_{i\in I}(SN_i) = S(\bigcap_{i\in I} N_i)$.

Proof. Using Theorem 5.4 (ii), we may assume that $\bigcap_{i\in I} N_i = 1$ and we need to show that $\bigcap_{i\in I} SN_i = S$. Let $T = \bigcap_{i\in I} SN_i$; then clearly $T \geq S$ and so $TN_i = SN_i$. Hence $T/(T \cap N_i) \cong TN_i/N_i = SN_i/N_i \cong S/(S \cap N_i)$ is a π-group. Since $\bigcap_{i\in I}(T \cap N_i) = 1$, T is residually a π-group and so is itself a π-group. By the maximality of S we have T = S, as required.□

Lemma 5.6. Let $\{N_i : i \in I\}$ be a residual system of normal subgroups of finite index in the residually finite $\mathfrak{D}_\pi$-group G and, for each $i \in I$, let S_i/N_i be a Sylow π-subgroup of G/N_i such that, whenever $N_i \leq N_j$, S_iN_j is equal to S_j. Then $S = \bigcap_{i\in I} S_i$ is a Sylow π-subgroup of G.

Proof. Let T be a Sylow π-subgroup of G. Then, for each $i \in I$, TN_i/N_i is a Sylow π-subgroup of G/N_i, by Theorem 5.4 (ii), and so is conjugate to S_i/N_i. Now let A_i be the set of inner automorphisms of G/N_i which map TN_i/N_i onto S_i/N_i. Then each A_i is non-empty and if $N_i \leq N_j$ and $\theta \in A_i$, then $(TN_j/N_i)\theta = S_iN_j/N_i = S_j/N_i$ and so θ induces in G/N_j an automorphism from A_j. Therefore we can applv Theorem 4.17 to obtain a locally inner automorphism φ of G such that $(TN_i)\phi = S_i$, for each $i \in I$, and hence $S = \bigcap_{i\in I} S_i = \bigcap_{i\in I}(TN_i)\phi = (\bigcap_{i\in I} TN_i)\phi = T\phi$, by Lemma 5.5. Therefore S, being locally conjugate to T, is a Sylow π-subgroup of G. □

Kargapolov [58] considered the question of when the Sylow subgroups of a periodic FC-group G are conjugate. Theorem 5.2 shows that a Sylow π-subgroup of a $\mathfrak{D}_\pi$-group is L-pronormal and so Theorem 4.22 immediately gives a criterion for two Sylow π-subgroups to be conjugate.

Theorem 5.7. (Kargapolov [58]) Let S and T be Sylow π-subgroups of the $\mathfrak{D}_\pi$-group G. Then S and T are conjugate in G if and only if $|S:S\cap T|$ is finite. □

Theorem 5.8. (Kargapolov [58]) Let G be a $\mathfrak{D}_\pi$-group. Then the following are equivalent:

(a) The Sylow π-subgroups of G are conjugate in G.

(b) There are only finitely many Sylow π-subgroups of G.

(c) The Sylow π-subgroups of $G/O_\pi(G)$ are finite.

Proof. (b) ⇒ (c). Let S be a Sylow π-subgroup of G. Then $C\ell_G(S)$ is finite and so S/S_G is finite, by Lemma 4.21. It is clear that $S_G = O_\pi(G)$ and (c) is proved.

(c) ⇒ (a). Let S and T be Sylow π-subgroups of G; then $S/O_\pi(G)$ and $T/O_\pi(G)$ are finite Sylow π-subgroups of $G/O_\pi(G)$ and so are conjugate.

(a) ⇒ (b). Let S be a Sylow π-subgroup of G; then SZ/Z is a Sylow π-subgroup of the $\mathfrak{Y}$-group G/Z. If T_1/Z is a Sylow π-subgroup of G/Z, then again using Lemma 5.4, we have $T_1 = TZ$ where T is a Sylow π-subgroup of G. Now T is conjugate to S and so T_1/Z is conjugate to SZ/Z. Therefore $Lc\ell(SZ/Z) = C\ell(SZ/Z)$ and so, by Theorem 4.25 (ii), $C\ell(SZ)$ is finite. If $x^{-1}Sx$ is a conjugate of S, then $x^{-1}Sx$ is the unique Sylow π-subgroup of $x^{-1}SZx$ and hence $C\ell(S)$ is finite. □

We saw in the proof of Theorem 5.4(ii) how Lemma 5.3 could be used to check that a given subgroup is a Sylow π-subgroup and the dual result Lemma 5.6 could be used in a similar way. Another useful test for a subgroup being a Sylow π-subgroup is given in the first part of the following result on products of π-subgroups and π'-subgroups. As usual π' denotes the set of all primes not in π.

Lemma 5.9.(i) Let the locally finite group G be the product ST of a π-subgroup S and a π'-subgroup T. Then S is a Sylow π-subgroup of G (and T is a Sylow π'-subgroup of G).

(ii) Let S be a Sylow π-subgroup and T a Sylow π'-subgroup of the locally soluble periodic FC-group G. Then G = ST.

Proof. (i) Let S_1 be a Sylow π-subgroup of G containing S. Then $S_1 = S_1 \cap ST = S(S_1 \cap T) = S$.

(ii) Let $\{F_i : i \in I\}$ be a local system of finite normal subgroups of G. Then, for each $i \in I$, $S \cap F_i$ is a Sylow π-subgroup of F_i and $T \cap F_i$ is a Sylow π'-subgroup of F_i. Since F_i is a finite soluble group, $(S \cap F_i)(T \cap F_i) = F_i$. It follows that $ST \supseteq \bigcup_{i \in I}(S \cap F_i)(T \cap F_i) = \bigcup_{i \in I} F_i = G$ and so $G = ST$. □

Another useful result concerning products of subgroups in an FC-group but probably not in an arbitrary locally finite group is the following.

Lemma 5.10. *Let the periodic FC-group G be the product of two π-subgroups $G = ST$. Then G is also a π-group.*

Proof. Let $\{N_i/Z : i \in I\}$ be a residual system of normal subgroups of finite index in G/Z. For each $i \in I$, the finite group G/N_i is the product of the π-subgroups SN_i/N_i and TN_i/N_i and hence G/N_i is a π-group. It follows that G/Z is a π-group and it is sufficient to prove that each element of Z is a π-element. With the obvious notation, let $z = st \in Z$. Then $z = s^{-1}zs = ts$ so that s and t commute. Therefore $z^n = s^n t^n$ for all integers n and it is clear that z is a π-element. □

The next sequence of results is included specifically for the formation theory which will follow.

Lemma 5.11. *Let N and H be normal subgroups and S a Sylow π-subgroup of the $\mathfrak{D}_\pi$-group G. Then*

(i) $NH \cap NS = N(H \cap S)$,

(ii) $N_G(NH \cap NS) = NN_G(H \cap S) = (H \cap N)\, N_G(H \cap S)$.

Proof. (i). By Theorem 5.4(i), $H \cap S$ is a Sylow π-subgroup of H and, using Theorem 5.4(ii), it follows that $N(H \cap S)/N$ is a Sylow π-subgroup of NH/N.

Applying the two parts of Theorem 5.4 in the reverse order, we see that NS/N is a Sylow π-subgroup of G/N and hence $(NH \cap NS)/N$ is a Sylow π-subgroup of NH/N.

But it is obvious that $N(H \cap S) \leq NH \cap NS$ and so we have the required equality.

(ii) It is clear that

$$(H \cap N)N_G(H \cap S) \leqslant NN_G(H \cap S) \leqslant N_G(N(H \cap S)) = N_G(NH \cap NS),$$

using part (i).

To prove the reverse inequalities, let $x \in N_G(NH \cap NS)$ so that

$$NH \cap N(x^{-1}Sx) = x^{-1}(NH \cap NS)x = NH \cap NS$$

and hence

$$H \cap x^{-1}Sx \leqslant H \cap (NH \cap NS) = H \cap N(H \cap S) = (H \cap N)(H \cap S).$$

Now x is contained in a finite normal subgroup F of G and so $H \cap x^{-1}Sx = x^{-1}(H \cap S)x \leqslant F(H \cap S)$. Now $H \cap S$ and $H \cap x^{-1}Sx$ are both Sylow π-subgroups of H and are contained in the subgroup $K = (H \cap N)(H \cap S) \cap F(H \cap S)$ of H; therefore $H \cap S$ and $H \cap x^{-1}Sx$ are Sylow π-subgroups of K. Also $|H \cap S:(H \cap S) \cap (H \cap x^{-1}Sx)|$ is finite and so, by Theorem 5.7, $H \cap S$ and $x^{-1}(H \cap S)x$ are conjugate in K. Since $K \leqslant (H \cap N)(H \cap S)$, there is an element $h \in H \cap N$ such that $x^{-1}(H \cap S)x = h^{-1}(H \cap S)h$. Thus $xh^{-1} \in N_G(H \cap S)$ and so $x \in (H \cap N)N_G(H \cap S)$. □

Part (ii) of the above result is just a variation on the usual Frattini argument and, putting H = N, we obtain the more usual form.

Corollary 5.12. (Frattini - argument) Let N be a normal subgroup and S a Sylow π-subgroup of the $\mathfrak{D}_\pi$-group G. Then $NN_G(N \cap S) = G$. □

Lemma 5.13. Let H and N_i, $i \in I$, be normal subgroups and S a Sylow π-subgroup of the $\mathfrak{D}_\pi$-group G. Then

$$\bigcap_{i\in I} N_i N_G(H \cap S) = (\bigcap_{i\in I} N_i)N_G(H \cap S).$$

Proof. $\bigcap_{i\in I} N_i N_G(H \cap S) = \bigcap_{i\in I} N_G(N_i H \cap N_i S)$, by Lemma 5.11 (ii),

$$\leqslant N_G(\bigcap_{i\in I}(N_i H \cap N_i S)),$$

$= N_G(\bigcap_{i\in I}(N_iH) \cap \bigcap_{i\in I}(N_iS))$,

$= N_G(\bigcap_{i\in I}(N_iH) \cap (\bigcap_{i\in I}N_i)S)$, by Lemma 5.5,

$\leq N_G((\bigcap_{i\in I}N_i)H \cap (\bigcap_{i\in I}N_i)S)$,

$= (\bigcap_{i\in I}N_i)N_G(H \cap S)$, by Lemma 5.11(ii).

The reverse inequality is clear and so we have the required equality. □

Lemma 5.14. *Let N be a normal subgroup of the locally soluble periodic FC-group G and let S_π, $S_{\pi'}$ be Sylow π- and π′-subgroups of G, respectively. Then $S_\pi \cap N_G(N \cap S_{\pi'})$ is a Sylow π-subgroup of $N_G(N \cap S_{\pi'})$.*

Proof. It is clear that $S_{\pi'} \leq N_G(N \cap S_{\pi'})$ and, by Lemma 5.9(ii), $G = S_\pi S_{\pi'}$. Therefore $N_G(N \cap S_{\pi'}) = (S_\pi \cap N_G(N \cap S_{\pi'}))S_{\pi'}$ and it follows from Lemma 5.9(i) that $S_\pi \cap N_G(N \cap S_{\pi'})$ is a Sylow π-subgroup of $N_G(N \cap S_{\pi'})$. □

We now introduce certain sets of Sylow subgroups of a group which fit together in a rather nice way. A *Sylow basis* of a group G is a set $\mathcal{S} = \{S_p\}$ of Sylow p-subgroups of G, one for each prime p, such that $\langle S_p : p \in \pi\rangle$ is a π-group, for each set of primes π. Other authors occasionally define a Sylow basis by the equivalent conditions given in the following result. We shall not in fact use this equivalence but include it for completeness.

Lemma 5.15. *Let $\mathcal{S} = \{S_p\}$ be a set of Sylow p-subgroups of the periodic FC-group G, one for each prime p. Then the following are equivalent:*

(a) *$\mathcal{S}$ is a Sylow basis of G.*

(b) *$S_pS_q = S_qS_p$, for all primes p,q.*

Proof. Suppose that $\mathcal{S}$ is a Sylow basis of G. Then $H = \langle S_p, S_q\rangle$ is a {p,q}-group. Let $\{F_i : i \in I\}$ be a local system of finite normal subgroups of H. Then $S_p \cap F_i$ is a Sylow p-subgroup of F_i and $S_q \cap F_i$ is a Sylow q-subgroup of F_i. Since F_i is a finite {p,q}-group, we have $F_i = (F_i \cap S_p)(F_i \cap S_q)$ and so $H = \bigcup_{i\in I}F_i \subseteq S_pS_q$. Hence $H = S_pS_q = S_qS_p$.

Conversely, suppose that $S_pS_q = S_qS_p$, for all primes p,q and let $K = \langle S_p : p \in \pi\rangle$. If F is a finite subgroup of K then there is a finite set of primes $\{p_1,\ldots,p_k\} \subseteq \pi$ such that $F \leq \langle S_{p_1},\ldots,S_{p_k}\rangle = S_{p_1}\ldots S_{p_k}$. By

Lemma 5.10, $S_{p_1} \dots S_{p_k}$ is a $\{p_1, \dots, p_k\}$-group and so F is a π-group. Hence K is a π-group, as required. □

It is immediate from our definition and the corresponding result for Sylow p-subgroups (Theorem 5.4) that Sylow bases behave well with respect to normal subgroups and factor groups.

Lemma 5.16. Let N be a normal subgroup and $\mathcal{S} = \{S_p\}$ a Sylow basis of the periodic FC-group G. Then

(i) $\mathcal{S} \cap N = \{S_p \cap N\}$ is a Sylow basis of N,

(ii) $\mathcal{S}\, N/N = \{S_p N/N\}$ is a Sylow basis of G/N. □

Lemma 5.17. If $\mathcal{S} = \{S_p\}$ is a Sylow basis of the periodic FC-group G then $S_\pi = \langle S_p : p \in \pi \rangle$ is a Sylow π-subgroup of G.
(We call S_π the Sylow π-subgroup of G associated with $\mathcal{S}$.)

Proof. Since $\mathcal{S} \cap F$ is a Sylow basis of F for each finite normal subgroup F of G, we see by considering orders that $\langle S_p \cap F : p \in \pi \rangle$ is a Sylow π-subgroup of F. Since $S_\pi \cap F \geqslant \langle S_p \cap F : p \in \pi \rangle$ and S_π is a π-group we must have $S_\pi \cap F = \langle S_p \cap F : p \in \pi \rangle$ and $S_\pi \cap F$ is a Sylow π-subgroup of F. Hence, by Lemma 5.3, S_π is a Sylow π-subgroup of G. □

The results given above will hold whenever the group G has a Sylow basis but we have so far said nothing about the existence of Sylow bases. Our aim will be to construct Sylow bases of G from Sylow bases of the finite normal subgroups of G or, dually, from the finite factor groups associated with a residual system of a residually finite periodic FC-group.

Lemma 5.18. Let $\{F_i : i \in I\}$ be a local system of finite normal subgroups of the periodic FC-group G and, for each $i \in I$, let $\mathcal{S}_i = \{S_p(i)\}$ be a Sylow basis of F_i such that, whenever $F_i \geqslant F_j$, $\mathcal{S}_i \cap F_j = \mathcal{S}_j$. If we define $S_p = \bigcup_{i \in I} S_p(i)$, then $\mathcal{S} = \{S_p\}$ is a Sylow basis of G.

Proof. By Lemma 5.3, each S_p is a Sylow p-subgroup of G. Any element of $\langle S_p : p \in \pi \rangle$ will be contained in a subgroup $\langle S_p \cap F_i : p \in \pi \rangle = \langle S_p(i) : p \in \pi \rangle$

and this is a π-group. Thus each element of $\langle S_p : p \in \pi \rangle$ is a π-element and so S is a Sylow basis of G. □

Lemma 5.19. Let $\{N_i : i \in I\}$ be a residual system of normal subgroups of finite index in the residually finite periodic FC-group G and, for each $i \in I$, let $S_i = \{S_p(i)/N_i\}$ be a Sylow basis of G/N_i such that, whenever $N_i \leq N_j$, $S_i N_j/N_j = S_j$. If we define $S_p = \bigcap_{i\in I} S_p(i)$, then $S = \{S_p\}$ is a Sylow basis of G.

Proof. By Lemma 5.6, each S_p is a Sylow p-subgroup of G. The subgroup $\langle S_p : p \in \pi \rangle$ is contained in $T_i = \langle S_p(i) : p \in \pi \rangle$ and since S_i is a Sylow basis of G/N_i, T_i/N_i is a π-group. Hence $\bigcap_{i\in I} T_i$ is a π-subgroup of G and hence $\langle S_p : p \in \pi \rangle$ is a π-group and S is a Sylow basis of G. □

To make use of either of these lemmas we need the existence of Sylow bases in finite groups. We are also concerned with the relation between different Sylow bases and we say that two Sylow bases $S = \{S_p\}$ and $T = \{T_p\}$ of a group G are *(locally) conjugate* in G if there is a (locally) inner automorphism ϕ of G such that $S_p\phi = T_p$, for each prime p. Our main theorem will then be built up by applying the following well known theorem of Hall ([55], p.665) in the finite normal subgroups of G.

Theorem 5.20. (Hall) A finite group G has a Sylow basis if and only if G is soluble and, in this case, any two Sylow bases of G are conjugate in G. □

To make use of Hall's Theorem we shall need to consider the inverse limit of finite sets rather than of finite groups as considered in Chapter 4.

An *inverse system* of sets is a family $\{X_i : i \in I\}$ of sets indexed by a directed set I together with mappings $\theta_{ji} : X_j \to X_i$, for each pair $i, j \in I$ with $i < j$, satisfying the conditions

(IS1) θ_{ii} is the identity mapping on X_i

(IS2) if $i < j < k$, then $\theta_{kj}\theta_{ji} = \theta_{ki} : X_k \to X_i$.

The *inverse limit* $\varprojlim X_i$ of this system (or the inverse limit of the family $\{X_i : i \in I\}$) is defined to be the subset of the cartesian product $\Pi_{i\in I} X_i$ consisting of those elements (x_i) such that, whenever $i < j$, $x_j\theta_{ji} = x_i$.

If $(x_i) \in \varprojlim X_i$, then the set of components $\{x_i : i \in I\}$ forms a *complete projection set* in the terminology of Kuroš ([66] , p.167). All our results are based on the following well-known theorem on inverse limits of finite sets which may be found in Bourbaki ([11] , p.202) or Kuroš ([66] ,p.167).

Theorem 5.21. If $\{X_i : i \in I\}$ is an inverse system of finite non-empty sets, then $\varprojlim X_i$ is non-empty; that is, there exist complete projection sets. □

This theorem will usually be applied when the X_i are sets of subgroups (or, below, sets of Sylow bases) of finite normal subgroups in a local system or of finite factor groups associated with a residual system.

Theorem 5.22. (Gol'berg [40], Stonehewer [93]) A periodic FC-group G has a Sylow basis if and only if G is locally soluble and, in this case, any two Sylow bases of G are locally conjugate in G.

Proof. If G has a Sylow basis then, by Lemma 5.16(i), each finite normal subgroup of G has a Sylow basis and hence, by Hall's Theorem 5.20, is soluble. Therefore G is locally soluble.

Conversely, suppose that G is locally soluble and let $\{F_i : i \in I\}$ be a local system of finite normal subgroups of G. Then each F_i has a Sylow basis. For each $i \in I$, let X_i be the finite non-empty set consisting of all Sylow bases of F_i. The index set I can be made into a directed set by defining $i < j$ to mean $F_i \leqslant F_j$. If $i < j$ and $\mathcal{S} \in X_j$, then $\mathcal{S} \cap F_i$ is a Sylow basis of F_i and we can define $\theta_{ji} : X_j \to X_i$ by $\mathcal{S}\theta_{ji} = \mathcal{S} \cap F_i$. These definitions make $\{X_i : i \in I\}$ into an inverse system of finite non-empty sets. By Theorem 5.21, the inverse limit is non-empty and so there is a set $\{\mathcal{S}_i : i \in I\}$ such that each $\mathcal{S}_i$ is a Sylow basis of F_i and, whenever $F_i \geqslant F_j$, $\mathcal{S}_i \cap F_j = \mathcal{S}_j$. If $\mathcal{S}_i = \{S_p(i)\}$ and we define $S_p = \bigcup_{i \in I} S_p(i)$, then it follows from Lemma 5.18 that $\mathcal{S} = \{S_p\}$ is a Sylow basis of G.

Now let $\mathcal{S} = \{S_p\}$ and $\mathcal{T} = \{T_p\}$ be two Sylow bases of the locally soluble periodic FC-group G. Then, for each $i \in I$, $\mathcal{S} \cap F_i$ and $\mathcal{T} \cap F_i$ are Sylow bases of F_i, by Lemma 5.16(i), and so, by Hall's Theorem 5.20, are conjugate in F_i. Let A_i be the set of automorphisms of F_i induced by inner automorphisms of G which map $S_p \cap F_i$ onto $T_p \cap F_i$, for each prime p. If $F_i \geqslant F_j$ and $\theta \in A_i$ then, for each prime p, $(S_p \cap F_j)\theta = (S_p \cap F_i)\theta \cap F_j = T_p \cap F_j$ and so the automorphism induced in F_j by θ is in A_j. We can therefore apply Theorem 4.16 to obtain a locally inner automorphism ϕ of G such that $(S_p \cap F_i)\phi = T_p \cap F_i$, for each prime p and each $i \in I$. Hence $S_p\phi = T_p$, for each prime p and so S and T are locally conjugate in G. □

Hall's Theorem 5.20 was originally given in terms of Sylow complement systems. A *Sylow complement system* $K = \{S_{p'}\}$ of a group G is defined to be a set of Sylow p′-subgroups of G, one for each prime p. It is clear from Lemma 5.17 that a Sylow basis of a periodic FC-group G determines a Sylow complement system of G. The following result shows that the correspondence is one to one and that the Sylow complement systems of G are also locally conjugate in G.

Theorem 5.23. Let $K = \{S_{p'}\}$ be a Sylow complement system of the periodic FC-group G. If we define $S_p = \bigcap_{q \neq p} S_{q'}$, then S_p is a Sylow p-subgroup of G, $S = \{S_p\}$ is a Sylow basis of G and $S_{p'}$ is the Sylow p′-subgroup of G associated with S.

Hence a periodic FC-group G has a Sylow complement system if and only if G is locally soluble and, in this case, any two Sylow complement systems of G are locally conjugate in G.

Proof. Let $\{F_i : i \in I\}$ be a local system of finite normal subgroups of G. For each $i \in I$ and for each prime p, $S_{p'} \cap F_i$ is a Sylow p′-subgroup of F_i, by Theorem 5.4(i), and so $K \cap F_i = \{S_{p'} \cap F_i\}$ is a Sylow complement system of F_i. By the known theorem on finite groups, $S_p \cap F_i = \bigcap_{q \neq p}(S_{q'} \cap F_i)$ is a Sylow p-subgroup of F_i and hence, by Lemma 5.3, $S_p = \bigcup_{i \in I}(S_p \cap F_i)$ is a Sylow p-subgroup of G.

It is also known that $S \cap F_i = \{S_p \cap F_i\}$ is a Sylow basis of F_i and hence, by Lemma 5.18, S is a Sylow basis of G. The remaining statements are clear. □

In infinite groups it is more usual to use Sylow bases rather than complement systems as, in general, complement systems do not give rise to Sylow bases in the way indicated above and will not usually have good conjugacy properties.

A simple example of this is given by letting A be the direct product of cyclic groups $\langle a_p \rangle$ of order p, one for each odd prime p, and letting G be the split extension of A by a group $\langle x \rangle$ of order 2 such that $x^{-1}ax = a^{-1}$, for all $a \in A$. For each odd prime p, G has a Sylow p′-subgroup $S_{p'} = A_{p'} \langle a_p^{-1} x a_p \rangle$ and $A = S_{2'}$ is a Sylow 2′-subgroup. However $\bigcap_{p\neq 2} S_{p'} = 1$ which is clearly not a Sylow 2-subgroup of G, so the complement system $\{S_{p'}\}$ does not give rise to a Sylow basis. Also G has a complement system $\{T_{p'}\}$ in which $T_{2'} = A$ and, for p odd, $T_{p'} = A_{p'}\langle x \rangle$. Since $\bigcap_{p\neq 2} T_{p'} = \langle x \rangle$, there is no automorphism ϕ of G such that $S_{p'}\phi = T_{p'}$, for all primes p.

The relationship between complement systems and Sylow bases in periodic FC-groups does make the proof of the following useful result very much easier.

Lemma 5.24. Let H be a subgroup of the locally soluble periodic FC-group G. If $\mathcal{T} = \{T_p\}$ is a Sylow basis of H, then there is a Sylow basis $\mathcal{S} = \{S_p\}$ of G such that $\mathcal{S} \cap H = \mathcal{T}$.

Proof. Let $T_{p'}$ be the Sylow p′-subgroup of H associated with $\mathcal{T}$. For each prime p, let $S_{p'}$ be a Sylow p′-subgroup of G containing $T_{p'}$. Then $\{S_{p'}\}$ forms a Sylow complement system of G. If $S_p = \bigcap_{q\neq p} S_{q'}$, then $\mathcal{S} = \{S_p\}$ is a Sylow basis of G and, for each prime p, $S_p \cap H = \bigcap_{q\neq p} S_{q'} \cap H = \bigcap_{q\neq p} T_{q'} = T_p$, as required. □

It is also possible to give a criterion for the Sylow bases of a locally soluble periodic FC-group to be conjugate rather than locally conjugate but discussion of this question will be rather simpler once the basis normalizers have been dealt with in the next chapter (see Corollary 6.31).

We complete this chapter by considering complementation results which are related to the Schur-Zassenhaus Theorem.

Theorem 5.25. (Černikov [14]) Let G be a periodic FC-group with a normal π-subgroup N such that G/N is a π′-group. Then N has a complement in G and all such complements are locally conjugate in G.

Proof. By the Schur-Zassenhaus Theorem for finite groups ([55] , p.126), G is a $\mathfrak{D}_{\pi'}$-group and so the Sylow π′-subgroups of G are locally conjugate.

by Theorem 5.2. If S is a Sylow π'-subgroup of G, then by Theorem 5.4(iii), SN = G and clearly $S \cap N = 1$ so that S is a complement to N in G. Conversely, every complement is a Sylow π'-subgroup and so the complements are locally conjugate in G. □

The usual proof of the Schur-Zassenhaus Theorem for finite groups reduces to the case in which N is abelian and uses cohomology theory to prove the existence and conjugacy of the complements in this case ([55] , p.122). This latter part of the proof was generalized by Gaschütz ([55] , p.121) as follows:

Theorem 5.26 (Gaschütz) Let A be an abelian normal subgroup of the group G and let B be a subgroup containing A such that $|G:B| = k$ and each element of A has a unique k th root. [e.g. A is a π-group and k a π'-number.]

(i) If A has a complement in B, then A has a complement in G.

(ii) If K_1 and K_2 are complements of A in G such that $K_1 \cap B = K_2 \cap B$ then K_1 and K_2 are conjugate in G. □

Possibly the most interesting question concerning FC-groups at the present time is whether this theorem can be generalized completely to periodic FC-groups or, if not, the extent to which it can be generalized. We ask the question more specifically for the case in which B is a Sylow p-subgroup of G and, while hoping to be proved wrong, phrase the question in the following pessimistic form.

Question 5A. Is there a periodic FC-group G with a Sylow p-subgroup P and normal p-subgroup A such that A has a complement in P but does not have a complement in G?

My pessimism about the probable answer to this question is not shared by everyone and D. Holt has proved a number of positive results. We describe one of these results which gives another interesting type of inverse limit argument.

First we introduce some notation: for any group X, we define $Q^p(X) = X'O^p(X)$ so that $Q^p(X)$ is the smallest normal subgroup of X such that $X/Q^p(X)$ is an abelian p-group.

Theorem 5.27. (Holt) Let P be a Sylow p-subgroup and A an abelian normal p-subgroup of the periodic FC-group G and let $Q/A = Q^p(G/A)$. If A has a complement in P, then A has a complement in Q.

In particular, if G/A is p-perfect then A has a complement in G.

Proof. Let $\{F_i/A : i \in I\}$ be a local system of Q/A consisting of finite normal subgroups of G/A. For each $i \in I$, we can choose a finite normal subgroup F_i^*/A of G/A containing F_i/A such that $F_i/A \leq Q^p(F_i^*/A)$ and, whenever $F_i \leq F_j$, we have $F_i^* \leq F_j^*$. [Suppose F_i^* has been defined for each $F_i < F_j$ and let $F^* = \langle F_i^* : F_i < F_j \rangle$. Then there is a finite normal subgroup F/A such that $F_jF^*/A \cap (G/A)' \leq (F/A)'$ and we can define $F_j^* = F_jF^*F$.]

For each $i \in I$, let X_i be the set of complements to A in F_i which are restrictions of complements to A in F_i^*. Since F_i^*/A is finite, it follows from Theorem 5.26 that A has a complement in F_i^* and so each X_i is non-empty. Now let $C_i = C_A(F_i^*)$; since F_i^* is abelian-by-finite and hence centre-by-finite it follows that F_i^*/C_i is finite and hence A/C_i has finitely many complements in F_i^*/C_i. We show that the set X_i is finite by showing that for each complement K/C_i of A/C_i in F_i^*/C_i, the complements of A in F_i^* contained in K all have the same intersection with F_i. Suppose then that K_1 and K_2 are complements of A in F_i^* such that K_1 and K_2 are contained in K. Then $K = K_1C_i = K_2C_i$. For each $k \in K/C_i$, let k_1 and k_2 be the inverse images of k in K_1 and K_2, respectively. Since C_i is central in K, the mapping $\sigma : K/C_i \to C_i$ defined by $k\sigma = k_1k_2^{-1}$ is a homomorphism. The image of σ is an abelian p-group and so $\mathrm{Ker}\sigma \geq Q^p(K/C_i^*) \geq (K \cap F_i)/C_i^*$. That is, for each $k \in (K \cap F_i)/C_i^*$, $k_1 = k_2$ and hence $K_1 \cap F_i = K_2 \cap F_i$. The sets X_i are therefore each finite and non-empty.

The index set I is made into a directed set by defining $i \prec j$ if and only if $F_i \leq F_j$. Whenever $i \prec j$, we define a map $\theta_{ji} : X_j \to X_i$ by $K_j\theta_{ji} = K_j \cap F_i$. With these definitions, the sets X_i form an inverse system of finite non-empty sets and hence $\varprojlim X_i$ is non-empty. Therefore there is a set $\{K_i : i \in I\}$, where each K_i is a complement to A in F_i and whenever $F_i \leq F_j$, we have $K_j \cap F_i = K_i$. It is then clear that $K = \bigcup_{i\in I} K_i$ is a complement to A in $\bigcup_{i \in I} F_i = Q$. □

Having obtained a complement to A in Q, Holt is then able to obtain a complement to A in G if additional assumptions are made about Q, essentially saying that Q is a large subgroup of G. These results are rather more technical and require some cohomological methods. We simply state the results.

Theorem 5.28. (Holt) With the notation of Theorem 5.27 if either G/Q is finite or $C_A(Q)$ is finite, then A has a complement in G. □

If A does have complement in G then part (ii) of Gaschütz's Theorem is easily extended.

Theorem 5.29. Let P be a Sylow p-subgroup and A an abelian normal p-subgroup of the periodic FC-group G. If K_1 and K_2 are complements to A in G such that $K_1 \cap P = K_2 \cap P$, then K_1 and K_2 are locally conjugate in G.

Proof. Let $\{F_i/A: i \in I\}$ be a local system of finite normal subgroups of G/A. Then $K_1 \cap F_i$ and $K_2 \cap F_i$ are finite complements to A in F_i. Let $L_i = \langle K_1 \cap F_i, K_2 \cap F_i \rangle^G$; then $\{L_i : i \in I\}$ is a local system of finite normal subgroups of $L = \langle K_1, K_2 \rangle^G$. Also $L_i = L_i \cap A(K_1 \cap F_i) = (L_i \cap A)(K_1 \cap F_i)$ and so $K_1 \cap F_i$ and $K_2 \cap F_i$ are complements to $A \cap F_i$ in L_i containing $K_1 \cap F_i \cap P$. By Theorem 5.26, $K_1 \cap F_i$ and $K_2 \cap F_i$ are conjugate in L_i and a straightforward application of Theorem 4.16 shows that K_1 and K_2 are locally conjugate in L and hence in G. □

It should also be noted that the existence of complements in G follows easily if the complements in P are locally conjugate.

Theorem 5.30. Let P be a Sylow p-subgroup and A an abelian normal p-subgroup of the periodic FC-group G. If A has a complement in P and all such complements are locally conjugate in P, then A has a complement in G and these complements are locally conjugate in G.

Proof. Let $\{F_i : i \in I\}$ be a local system of finite normal subgroups of G. Then $|F_iP:P|$ is finite and prime to p so that, by Gaschütz's Theorem, A has complements in F_iP. By Theorem 5.29, these complements are locally conjugate and so if K is a fixed complement to A in P we can choose a complement L_i to A in F_iP such that $L_i \geq K$. Then $(L_i \cap F_iA)A = (L_iA \cap F_iA) =$

$= F_iP \cap F_iA = F_iA$ so that $L_i \cap F_iA$ is a finite complement to A in F_iA. Now let

$$F_i^* = \langle L_j \cap F_jA:\ F_j \leq F_i \rangle^G ;$$

Then F_i^* is a finite normal subgroup of G contained in F_iA and $\{F_i^* : i \in I\}$ is a local system of finite normal subgroups. Let X_i be the set of complements to $A \cap F_i^*$ in F_i^* which contain $K \cap F_i^*$. Then X_i is finite and is non-empty since $(L_i \cap F_i^*)(A \cap F_i^*) = (L_i \cap F_iA)A \cap F_i^* = F_iA \cap F_i^* = F_i^*$ so that $L_i \cap F_i^* \in X_i$. A simple inverse limit argument provides a complement L to $A \cap G^*$ in $G^* = \langle F_i^* : i \in I \rangle$ containing K. It is clear that L is also a complement to A in G, since $G^*A = G$. □

6 Formations and fitting classes

The dual concepts of formation and Fitting class were introduced in finite soluble groups by Gaschütz [38] and by Fischer, Gaschütz and Hartley [33]. These concepts lead to conjugacy classes of subgroups which behave well with respect to factor groups and normal subgroups, respectively. The properties of these subgroups in finite soluble groups lead us naturally to constructions via inverse limit arguments in locally soluble FC-groups. For formation theory we are led to consider the finite factor groups and for Fitting classes the finite normal subgroups.

We begin by considering formations and largely follow the methods of [99] together with some of the ideas used by Gardiner, Hartley and Tomkinson in [37]. The important change to the original approach is that we make much greater use of the fact that chief factors of FC-groups are finite to make a definition via formations of finite soluble groups. One reason for this approach is that it highlights the main problem about our final result, the question of interpreting it in terms of formations of finite soluble groups.

The results could be given in a more general form by using a preformation function as in [37] but this does not seem to gain a great deal here and might obscure the main points which we wish to illustrate.

Throughout this chapter the class of locally soluble periodic FC-groups will be denoted by $\mathfrak{S}$ and, if $\mathfrak{X}$ is any class of groups, $\mathfrak{X}^*$ is used to denote the class of finite $\mathfrak{X}$-groups. Thus $\mathfrak{S}^*$ is the class of finite soluble groups. As $\mathfrak{F}$ will be used here to denote a formation we shall of course avoid its earlier use for the class of finite groups. If π is a set of primes, we use $\mathfrak{S}_\pi$ to denote the class of π-groups which are in $\mathfrak{S}$.

Recall that a *formation of finite soluble groups* is a subclass $\mathfrak{E}$ of $\mathfrak{S}^*$, such that

(F*1) $\mathfrak{E}$ is Q-closed; i.e., if $G \in \mathfrak{E}$ and $N \triangleleft G$, then $G/N \in \mathfrak{E}$; and

(F*2) $\mathfrak{E}$ is R_0-closed; i.e., if $G/N_1 \in \mathfrak{E}$ and $G/N_2 \in \mathfrak{E}$, then $G/(N_1 \cap N_2) \in \mathfrak{E}$.

We shall require an additional condition relating to the occurrence of $\mathfrak{E}$-groups as automorphism groups induced on chief factors in an FC-group. If

G is an $\mathfrak{S}$-group, we define the $(\mathfrak{E},p)$-*centralizer* of G to be

$$C_G(\mathfrak{E},p) = \bigcap\{C_G(H/K) : H/K \text{ is a p-chief factor of } G \text{ and } G/C_G(H/K) \in \mathfrak{E}\} .$$

The formation $\mathfrak{E}$ of finite soluble groups is said to be FC-*solid* if, for each p-chief factor U/V of an $\mathfrak{S}$-group G with $C_G(U/V) \geq C_G(\mathfrak{E},p)$, we have $G/C_G(U/V) \in \mathfrak{E}$.

Of course, a sufficient condition for a formation $\mathfrak{E}$ to be FC-solid is that every finite factor group of a $(R\,\mathfrak{E} \cap \mathfrak{S})$ - group is an $\mathfrak{E}$-group. This condition is, in fact, satisfied by most of the well known formations.

Lemma 6.1. If $\mathfrak{E}$ is an S_n-closed formation of finite soluble groups, then $\mathfrak{E}$ is FC-solid.

Proof. Let G/N be a finite factor group of the $(R\,\mathfrak{E} \cap \mathfrak{S})$ - group G. There is a finite normal subgroup X such that XN = G, and there is a normal subgroup H such that $G/H \in R_0\mathfrak{E} = \mathfrak{E}$ and $H \cap X = 1$. Thus $X \cong XH/H \in S_n\mathfrak{E} = \mathfrak{E}$ and hence $G/N \cong X/(X \cap N) \in Q\mathfrak{E} = \mathfrak{E}$. □

If π is a non-empty set of primes, a *formation function* $\mathfrak{f}$ on π associates with each $p \in \pi$ an FC-solid formation $\mathfrak{f}(p)$ of finite soluble groups. The *saturated formation* $\mathfrak{F} = \mathfrak{F}(\mathfrak{f})$ of $\mathfrak{S}$-groups defined by the (FC-solid) formation function $\mathfrak{f}$ consists of those $\mathfrak{S}_\pi$-groups G in which the group of automorphisms induced by G in each p-chief factor is an $\mathfrak{f}(p)$-group; that is, $G/C_G(H/K) \in \mathfrak{f}(p)$, whenever H/K is a p-chief factor of G. Using Theorem 1.18, we see that in an $\mathfrak{F}$-group G, $O_{p',p}(G) = C_G(\mathfrak{f}(p),p)$ and so each finite factor group of G is in the class $\mathfrak{S}^*_\pi \cap \bigcap_{p\in\pi} \mathfrak{S}^*_{p'}\mathfrak{S}^*_p\,\mathfrak{f}(p)$, which is the saturated formation of finite soluble groups defined by the formation function $\mathfrak{f}$. The following lemma is obvious but important in that it allows us to use known results about finite soluble groups.

Lemma 6.2. If $\mathfrak{f}$ is a formation function on a set of primes π , then $\mathfrak{F}(\mathfrak{f})^*$ is the saturated formation of $\mathfrak{S}^*$-groups defined by $\mathfrak{f}$. □

A *formation* of $\mathfrak{S}$-groups is a subclass $\mathfrak{F}$ of $\mathfrak{S}$ satisfying

(F1) $\mathcal{F}$ is Q-closed, and

(F2) $R\mathcal{F} \cap \mathfrak{S} = \mathcal{F}$.

These two conditions ensure that any $\mathfrak{S}$-group G has a unique normal subgroup $G^{\mathcal{F}}$ such that $G/N \in \mathcal{F}$ if and only if $N \geq G^{\mathcal{F}}$. The subgroup $G^{\mathcal{F}}$ is called the *$\mathcal{F}$-residual* of G.

Lemma 6.3. If $\mathfrak{f}$ is a formation function on a set of primes π, then $\mathcal{F}(\mathfrak{f})$ is a formation of $\mathfrak{S}$-groups.

Proof. Let $G \in \mathcal{F}(\mathfrak{f})$ and let $N \lhd G$. Certainly G/N is a π-group. If $(H/N)/(K/N)$ is a p-chief factor of G/N, then H/K is a p-chief factor of G and the group of automorphisms induced in $(H/N)/(K/N)$ by G/N is isomorphic to $G/C_G(H/K)$, an $\mathfrak{f}(p)$-group.

Now let G be an $\mathfrak{S}$-group with normal subgroups $H_i, i \in I$, such that $G/H_i \in \mathcal{F}(\mathfrak{f})$ and $\bigcap_{i \in I} H_i = 1$. Again it is clear that G is a π-group. Let H/K be a p-chief factor of G; then there is a finite normal subgroup X of G such that XK = H and so $H/K \overset{G}{\cong} X/(X \cap K)$. There is a finite intersection $H_1 \cap \ldots \cap H_r$ such that $(H_1 \cap \ldots \cap H_r) \cap X = 1$. By considering the finite chain

$$X \geq X \cap H_1 \geq X \cap H_1 \cap H_2 \geq \ldots \geq X \cap H_1 \cap \ldots \cap H_r = 1,$$

we see that $X/(X \cap K)$ is G-isomorphic to some p-chief factor L/M of G with

$$X \cap H_1 \cap \ldots \cap H_{s-1} \geq L > M \geq X \cap H_1 \cap \ldots \cap H_s.$$

But then $L/M \cong H_sL/H_sM$, a p-chief factor above H_s. Thus

$$G/C_G(H/K) = G/C_G(H_sL/H_sM) \in \mathfrak{f}(p).$$

and so $G \in \mathcal{F}(\mathfrak{f})$. □

Lemma 6.4. If $\mathfrak{f}$ is a formation function on π and $\mathfrak{f}_1(p) = \mathcal{F}(\mathfrak{f}) \cap \mathfrak{f}(p)$, then $\mathfrak{f}_1(p)$ is an FC-solid formation of $\mathfrak{S}^*$-groups.

Proof. It is clear from Lemma 6.2 that $\mathfrak{f}_1(p)$ is a formation of finite soluble groups. To show that $\mathfrak{f}_1(p)$ is FC-solid, let G be an $\mathfrak{S}$-group and let

U/V be a p-chief factor of G with $C_G(U/V) \geq C = C_G(\mathfrak{f}_1(p),p)$. Now $G/C \in R\mathfrak{f}_1(p) \cap \mathfrak{S} \subseteq R\mathfrak{F}(\mathfrak{f}) \cap \mathfrak{S} = \mathfrak{F}(\mathfrak{f})$ and so, since $\mathfrak{F}(\mathfrak{f})$ is also Q-closed, we have $G/C_G(U/V) \in \mathfrak{F}(\mathfrak{f})$. Also, since $\mathfrak{f}(p)$ is FC-solid, $G/C_G(U/V) \in \mathfrak{f}(p)$. Thus $G/C_G(U/V) \in \mathfrak{f}_1(p)$, as required. □

It is clear that $\mathfrak{F}(\mathfrak{f}_1) = \mathfrak{F}(\mathfrak{f})$ and so, from now on, we shall assume that $\mathfrak{f}(p) \subseteq \mathfrak{F}(\mathfrak{f})$, for all $p \in \pi$; that is, $\mathfrak{f}$ is an *integrated formation function*. Lemmas 6.2 and 6.3 are the results we shall use in proving our main theorems but to interpret the final results we need to give a characterization of $\mathfrak{F}(\mathfrak{f})$ in terms of formations of finite soluble groups.

Lemma 6.5. *Let $\mathfrak{f}$ be a formation function on π. Then the $\mathfrak{S}$-group G is an $\mathfrak{F}(\mathfrak{f})$-group if and only if G is a π-group and $G/Z(G) \in R(\mathfrak{F}(\mathfrak{f})^*)$.*

Proof. If G/Z is residually an $\mathfrak{F}(\mathfrak{f})^*$-group, then $G/Z \in R\mathfrak{F}(\mathfrak{f}) \cap \mathfrak{S} = \mathfrak{F}(\mathfrak{f})$, by Lemma 6.3. Thus G induces an $\mathfrak{f}(p)$-group of automorphisms in each p-chief factor and so G is an $\mathfrak{F}(\mathfrak{f})$-group.

Conversely, if $G \in \mathfrak{F}(\mathfrak{f})$ then each finite factor group is in $\mathfrak{F}(\mathfrak{f})^*$, by the Q-closure of $\mathfrak{F}(\mathfrak{f})$, and hence $G/Z(G) \in R\mathfrak{F}(\mathfrak{f})^*$. □

The S_n-closed saturated formations can be described completely in terms of saturated formations of finite soluble groups.

Lemma 6.6. *If $\mathfrak{F}(\mathfrak{f})$ is S_n-closed, then $\mathfrak{F}(\mathfrak{f}) = L(\mathfrak{F}(\mathfrak{f})^*) \cap \mathfrak{S}$. Conversely, if $\mathfrak{E}$ is any S_n-closed saturated formation of finite soluble groups, then $L\mathfrak{E} \cap \mathfrak{S}$ is a saturated formation of $\mathfrak{S}$-groups.*

Proof. (i) Let $G \in \mathfrak{F}(\mathfrak{f})$; then each finite normal subgroup is an $\mathfrak{F}(\mathfrak{f})^*$-group and so $G \in L(\mathfrak{F}(\mathfrak{f})^*) \cap \mathfrak{S}$. Conversely, if $G \in L(\mathfrak{F}(\mathfrak{f})^*) \cap \mathfrak{S}$, then each finite factor group of G is an $\mathfrak{F}(\mathfrak{f})^*$-group and so G/Z(G) is residually an $\mathfrak{F}(\mathfrak{f})^*$-group. By Lemma 6.5, $G \in \mathfrak{F}(\mathfrak{f})$.

(ii) Let $\mathfrak{E}$ be an S_n-closed saturated formation of finite soluble groups defined by the formation function $\mathfrak{f}$. Let $\mathfrak{f}_1(p) = \mathfrak{S}^*_p\,\mathfrak{f}(p)$; then $\mathfrak{E}$ can be defined by the formation function $\mathfrak{f}_1$ and we claim that $\mathfrak{f}_1(p)$ is S_n-closed and so is FC-solid, by Lemma 6.1. Let $G \in \mathfrak{S}^*_p\,\mathfrak{f}(p)$ and let $H \triangleleft G$. The group $W = C_p \text{ wr } G \in \mathfrak{S}^*_p\,\mathfrak{f}(p) \subseteq \mathfrak{E}$. Now W = BG, where B is the base group and $BH \in S_n \mathfrak{E} = \mathfrak{E}$. Thus $BH \in \mathfrak{S}^*_p, \mathfrak{S}^*_p f(p)$, but clearly $O_{p'}(BH) = 1$ and so $BH \in \mathfrak{S}^*_p f(p)$ and hence $H \cong BH/B \in \mathfrak{S}^*_p f(p)$.

It now follows from (i) and Lemma 6.2 that $\mathfrak{F}(f_1) = L\mathfrak{E} \cap \mathfrak{S}$. □

This result immediately gives us a number of examples of saturated formations of $\mathfrak{S}$-groups. The finite cases may be found in [55], pp. 697-698.

(1) If $f(p) = \{1\}$, for each prime p, then $\mathfrak{F}(f)$ is the class $L\mathfrak{N} \cap \mathfrak{S}$ of locally nilpotent periodic FC-groups.

(2) If $f(p)$ is the class of finite abelian groups of exponent dividing p-1, then $\mathfrak{F}(f)$ is the class $L\mathfrak{U} \cap \mathfrak{S}$ of locally supersoluble periodic FC-groups.

(3) If, for each prime p, $f(p)$ is the class of finite soluble groups of nilpotent length k-1, then $\mathfrak{F}(f)$ is the class $(L\mathfrak{N})^k \cap \mathfrak{S}$ of locally soluble periodic FC-groups of locally nilpotent length k.

(4) If, for each prime p, $f(p)$ is the class of finite abelian groups, then $\mathfrak{F}(f)$ is the class $(L\mathfrak{N})\mathfrak{A} \cap \mathfrak{S}$ of periodic FC-groups with locally nilpotent derived subgroups.

(5) Let p be a fixed prime; let $f(p) = \{1\}$ and, for each $q \neq p$, let $f(q)$ be the class of finite soluble groups. Then $\mathfrak{F}(f)$ is the class of locally p-nilpotent periodic FC-groups.

(6) Let p be a fixed prime; let $f(p) = \mathfrak{S}^*_{p'}$ and, for each $q \neq p$. let $f(p) = \mathfrak{S}^*$. Then $\mathfrak{F}(f)$ is the class of locally soluble periodic FC-groups of p-length one; that is, $G/O_{p'p}(G)$ is a p'-group.

For the formations which are not S_n-closed we can not give a simple characterization as in Lemma 6.6. That given in Lemma 6.5, although useful for proving theorems, tells us little about the classes that can occur as we have no simple test for a formation being FC-solid. This area lacks examples and we do not even have an example of a non-solid formation.

<u>Question 6A</u>. Is there a formation of finite soluble groups which is not FC-solid?

The same saturated formation may be defined by two different formation functions. For example, if we let $f(p)$ be the class of finite p-groups for each prime p, then $\mathfrak{F}(f)$ is the class of locally nilpotent periodic FC-groups which was obtained in example (1) above. However, there is a close relationship between different formation functions defining the same saturated formation.

Lemma 6.7. Let $\mathfrak{f}_1$ and $\mathfrak{f}_2$ be two formation functions on the set π such that $\mathfrak{F}(\mathfrak{f}_1) = \mathfrak{F}(\mathfrak{f}_2)$. Then $\mathfrak{S}^*_p \mathfrak{f}_1(p) = \mathfrak{S}^*_p \mathfrak{f}_2(p)$, for each $p \in \pi$.

Proof. We show that $\mathfrak{f}_1(p) \subseteq \mathfrak{S}^*_p \mathfrak{f}_2(p)$. Then, by symmetry, we also have $\mathfrak{f}_2(p) \subseteq \mathfrak{S}^*_p \mathfrak{f}_1(p)$ and so the result follows.

Let $G \in \mathfrak{f}_1(p)$; then C_p wr $G \in \mathfrak{S}^*_p \mathfrak{f}_1(p) \subseteq \mathfrak{F}^*(\mathfrak{f}_1) = \mathfrak{F}^*(\mathfrak{f}_2) \subseteq \mathfrak{S}^*_{p'} \mathfrak{S}^*_p \mathfrak{f}_2(p)$. But $O_{p'}(C_p \text{ wr } G) = 1$ and so C_p wr $G \in \mathfrak{S}^*_p \mathfrak{f}_2(p)$. Hence $G \in Q(\mathfrak{S}^*_p \mathfrak{f}_2(p) = \mathfrak{S}^*_p \mathfrak{f}_2(p)$. □

Since $O_p(G/C_G(H/K)) = 1$ for any p-chief factor H/K of an $\mathfrak{S}$-group G it follows immediately from Lemma 6.7 that $G/C_G(H/K) \in \mathfrak{f}_1(p)$ if and only if $G/C_G(H/K) \in \mathfrak{f}_2(p)$. Therefore if $\mathfrak{F} = \mathfrak{F}(\mathfrak{f})$ is a saturated formation of $\mathfrak{S}$-groups, the $(\mathfrak{f}(p),p)$-centralizer of an $\mathfrak{S}$-group G will depend only on the saturated formation $\mathfrak{F}$ and not on the particular formation function $\mathfrak{f}$ used to define $\mathfrak{F}$.

The saturated formation $\mathfrak{F}$ will now be assumed to be defined by the integrated formation function $\mathfrak{f}$ on the set π and, for each $p \in \pi$, C_p will denote the $(\mathfrak{f}(p),p)$-centralizer of the $\mathfrak{S}$-group G. If $S = \{S_p\}$ is a Sylow basis of G then we define the $\mathfrak{F}$-*normalizer* of G associated with S to be

$$D = S_\pi \cap \bigcap_{p\in\pi} N_G(C_p \cap S_{p'}),$$

where S_π and $S_{p'}$ denote the Sylow π- and p'-subgroups of G associated with S.

In particular, if $\mathfrak{F}$ is the class of locally nilpotent periodic FC-groups then, for each prime p, $C_p = G$ and so $D = \bigcap_p N_G(S_{p'}) = \bigcap_p N_G(S_p)$ is a *basis normalizer*.

The local conjugacy of the Sylow bases of G (Theorem 5.22) leads immediately to the first part of the following result.

Theorem 6.8. Let $\mathfrak{F}$ be a saturated formation of $\mathfrak{S}$-groups. Then

(i) any two $\mathfrak{F}$-normalizers of the $\mathfrak{S}$-group G are locally conjugate in G,

(ii) if G is an $\mathfrak{F}$-group, then the $\mathfrak{F}$-normalizers of G coincide with G.

Proof. (ii) If $G \in \mathfrak{F}$, then $C_p = O_{p'p}(G)$ and so $C_p \cap S_{p'}$ is the unique normal Sylow p'-subgroup of C_p. Hence $N_G(C_p \cap S_{p'}) = G$ and therefore $D = G$.□

Lemma 6.9. Let $\mathfrak{F}$ be a saturated formation of $\mathfrak{S}$-groups and let D be the $\mathfrak{F}$-normalizer of the $\mathfrak{S}$-group G associated with the Sylow basis $\mathcal{S}$. Then

(i) for each $p \in \pi$, $S_p \cap N_G(C_p \cap S_{p'}) = S_p \cap D$ is a Sylow p-subgroup of D,

(ii) $\mathcal{S}$ reduces into D; that is, $\mathcal{S} \cap D = \{S_p \cap D\}$ is a Sylow basis of D,

(iii) $D = \langle S_p \cap N_G(C_p \cap S_{p'}) : p \in \pi \rangle$.

Proof. (i) For each prime $q \neq p$, $S_p \leq S_{q'} \leq N_G(C_q \cap S_{q'})$ and so

$$S_p \cap N_G(C_p \cap S_{p'}) \leq S_\pi \cap \bigcap_{q\in\pi} N_G(C_q \cap S_{q'}) = D.$$

By Lemma 5.14, $S_p \cap N_G(C_p \cap S_{p'})$ is a Sylow p-subgroup of $N_G(C_p \cap S_{p'})$ and, since $D \leq N_G(C_p \cap S_{p'})$, $S_p \cap N_G(C_p \cap S_{p'})$ is therefore a Sylow p-subgroup of D. Thus $S_p \cap N_G(C_p \cap S_{p'}) = S_p \cap D$.

Parts (ii) and (iii) now follow immediately. □

There are two related properties of $\mathfrak{F}$-normalizers: their "homomorphism-invariance" and their cover-avoidance properties. The first of these will be an almost immediate consequence once we have established the second. If U/V is a chief factor of a group G and X a subgroup of G, we say that X *covers* U/V if $XV \geq U$ and hence $U/V \cong (X \cap U)/(X \cap V)$ and we say that X *avoids* U/V if $X \cap U \leq V$.

Lemma 6.10. If $S_{p'}$ is a Sylow p'-subgroup and H a normal subgroup of the $\mathfrak{S}$-group G then $N_G(H \cap S_{p'})$ avoids those p-chief factors U/V which are not centralized by H and covers all other chief factors of G.

Proof. By Lemma 5.11, we may assume that $V = 1$ so that U is a minimal normal subgroup of G.

(i) If U is a p'-group, then $U \leq S_{p'} \leq N_G(H \cap S_{p'})$.

(ii) If U is a p-group such that $[H,U] = 1$, then $U \leq C_G(H \cap S_{p'}) \leq N_G(H \cap S_{p'})$.

(iii) If U is a p-group such that $C_G(U) \not\geq H$, then let $M = C_G(U) \cap H < H$. Clearly $U \leq C_G(U)$ and if $H \cap U = 1$, we would have $H \leq C_G(U)$. Thus $U \leq C_G(U) \cap H = M$. Let L/M be a minimal normal subgroup of G/M contained in H/M. Then $L/M \cong LC_G(U)/C_G(U)$ and, since $O_p(G/C_G(U))$ is trivial, L/M must be a q-group, for some $q \neq p$. Therefore $L \leqslant M(H \cap S_{p'})$. Clearly

$$[U \cap N_G(H \cap S_{p'}),L] = [U \cap N_G(H \cap S_{p'}),(H \cap S_{p'})] \leqslant U \cap H \cap S_{p'} = 1$$

and so $C_G(L) \geqslant U \cap N_G(H \cap S_{p'})$. But $[L,U] \neq 1$ and so $C_U(L) < U$. Hence $C_U(L) = 1$ and so $U \cap N_G(H \cap S_{p'}) \leqslant U \cap C_G(L) = 1$. □

Lemma 6.11. *Let H and K be subgroups of the periodic FC-group G such that $K \leqslant H$ and each chief factor of G is either covered or avoided by H and by K. If H and K cover the same set of chief factors, then H = K.*

Proof. Let $\{F_i : i \in I\}$ be a local system of finite normal subgroups of G. Then by considering their orders, we see that $H \cap F_i = K \cap F_i$, for each $i \in I$, and hence $H = \bigcup_{i \in I}(H \cap F_i) = \bigcup_{i \in I}(K \cap F_i) = K$. □

Although this last lemma was extremely simple it should be noted that it is not true of groups in general. Hartley [50] has constructed examples of locally finite p-groups with proper subgroups which cover every chief factor.

If the saturated formation $\mathfrak{F}$ is defined by the formation function $\mathfrak{f}$, we say that a p-chief factor U/V of the $\mathfrak{S}$-group is *$\mathfrak{F}$-central* if $p \in \pi$ and $G/C_G(U/V) \in \mathfrak{f}(p)$. Otherwise U/V is *$\mathfrak{F}$-eccentric*. Again Lemma 6.7 shows that this definition is independent of the formation function used to define $\mathfrak{F}$. Because of the additional condition of being FC-solid which we imposed on the defining formations $\mathfrak{f}(p)$, the above condition is equivalent to $C_G(U/V) \geqslant C_p$. It is the use of this in the next few results which makes the restriction to FC-solid defining formations necessary.

Theorem 6.12. *Let $\mathfrak{F}$ be a saturated formation of $\mathfrak{S}$-groups and let D be the $\mathfrak{F}$-normalizer of $\mathfrak{S}$-group G associated with the Sylow basis $\mathcal{S}$. Then*

(i) *D covers each $\mathfrak{F}$-central chief factor of G and avoids each $\mathfrak{F}$-eccentric chief factor of G,*

(ii) *if $N \triangleleft G$, then DN/N is the $\mathfrak{F}$-normalizer of G/N associated with the Sylow basis $\mathcal{S}N/N$.*

(iii) *if $G/N \in \mathfrak{F}$, then DN = G.*

Proof. (i) If U/V is an $\mathfrak{F}$-central p-chief factor of G, then it follows from Lemma 6.10 that U/V is covered by $N_G(C_p \cap S_{p'})$ and so is covered by $S_p \cap N_G(C_p \cap S_{p'}) = S_p \cap D$.

If U/V is a π'-chief factor of G, then U/V is avoided by S_π and hence by D.

If U/V is an $\mathfrak{F}$-eccentric p-chief factor of G with $p \in \pi$, then it follows again from Lemma 6.10 that U/V is avoided by $N_G(C_p \cap S_{p'})$ and hence by D.

(ii) For each $p \in \pi$, let B_p/N be the $(f(p),p)$-centralizer of G/N and let $\bar{D}/N$ be the $\mathfrak{F}$-normalizer of G/N associated with $\mathcal{S}N/N$. Then

$$\bar{D} = \langle NS_p \cap N_G(B_p \cap NS_{p'}) : p \in \pi \rangle \leqslant \langle NS_p \cap N_G(NC_p \cap NS_{p'}) : p \in \pi \rangle .$$

Now, applying Lemma 5.14 and 5.11 in G/N, we see that $NS_p/N \cap N_G(NC_p \cap NS_{p'})/N$ is a Sylow p-subgroup of $NN_G(C_p \cap S_{p'})/N$. Also $S_p \cap N_G(C_p \cap S_{p'})$ is a Sylow p-subgroup of $N_G(C_p \cap S_{p'})$, by Lemma 5.14, and so $N(S_p \cap N_G(C_p \cap S_{p'}))$ is a Sylow p-subgroup of $NN_G(C_p \cap S_{p'})/N$.

Hence $NS_p \cap N_G(NC_p \cap NS_{p'}) = N(S_p \cap N_G(C_p \cap S_{p'}))$ and so

$$\begin{aligned} ND &= N \langle S_p \cap N_G(C_p \cap S_{p'}) : p \in \pi \rangle \\ &= \langle N(S_p \cap N_G(C_p \cap S_{p'})) : p \in \pi \rangle \\ &= \langle NS_p \cap N_G(NC_p \cap NS_{p'}) : p \in \pi \rangle \\ &\geqslant \bar{D}. \end{aligned}$$

But $\bar{D}/N$ and ND/N cover the same chief factors in G/N and hence, by Lemma 6.11, $\bar{D} = DN$.

(iii) follows immediately from (ii) and Theorem 6.8 (ii). □

Theorem 6.13. Let $\mathfrak{F}$ be a saturated formation of $\mathfrak{S}$-groups. Then the $\mathfrak{F}$-normalizers of an $\mathfrak{S}$-group are $\mathfrak{F}$-groups.

Proof. Let $(H_\alpha : \alpha < \rho)$ be an ascending chief series (Theorem 1.13) of the $\mathfrak{S}$-group G and let D be an $\mathfrak{F}$-normalizer of G. Then $(D \cap H_\alpha : \alpha < \rho)$ may be refined to give an ascending chief series $(U_\beta : \beta < \sigma)$ of D. By Theorem 1.13, it is sufficient to show that, for each p-chief factor $U_{\beta+1}/U_\beta$, we have $D/C_D(U_{\beta+1}/U_\beta) \in f(p)$. By the construction of the U_β, there is an $\alpha<\rho$ such that $D \cap H_\alpha \leqslant U_\beta < U_{\beta+1} \leqslant D \cap H_{\alpha+1}$. Clearly $H_{\alpha+1}/H_\alpha$ is a p-chief factor of G and D covers $H_{\alpha+1}/H_\alpha$ so that $H_{\alpha+1}/H_\alpha$ is $\mathfrak{F}$-central and so $C_G(H_{\alpha+1}/H_\alpha) \geqslant C_p$.

Now $G/C_p \in R\,\mathfrak{f}(p) \cap \mathfrak{S} \subseteq R\,\mathfrak{F} \cap \mathfrak{S} = \mathfrak{F}$ and so, by Theorem 6.12, $DC_p = G$. Therefore $DC_G(H_{\alpha+1}/H_\alpha) = G$ and so

$$D/C_D(H_{\alpha+1}/H_\alpha) \cong G/C_G(H_{\alpha+1}/H_\alpha) \in \mathfrak{f}(p) .$$

But $C_D(U_{\beta+1}/U_\beta) \geq C_D(H_{\alpha+1}/H_\alpha)$ and so we have $D/C_D(U_{\beta+1}/U_\beta) \in Q\mathfrak{f}(p) = \mathfrak{f}(p)$.□

For our main results we only require the following theorem in the finite case but it is no more difficult to prove for $\mathfrak{S}$-groups.

Theorem 6.14. Let $\mathfrak{F}$ be a saturated formation of $\mathfrak{S}$-groups and let G be an $\mathfrak{S}$-group with abelian $\mathfrak{F}$-residual A. Then

(i) there are no $\mathfrak{F}$-central chief factors of G below A,

(ii) A is complemented in G and the complements are the $\mathfrak{F}$-normalizers of G,

(iii) if H is an $\mathfrak{F}$-subgroup of G such that HA = G, then H is an $\mathfrak{F}$-normalizer of G. Hence the $\mathfrak{F}$-normalizers of G are maximal $\mathfrak{F}$-subgroups of G.

Proof. (i) Let C_p be the $(\mathfrak{f}(p),p)$-centralizer of G so that $C_p \geq A$ and $C_p/A \in \mathfrak{S}_{p'}\mathfrak{S}_p$. Let $L/A = O_{p'}(G/A)$. If U/V is a p-chief factor of G, then $O_p(G/C_G(U/V))$ is trivial and so $C_G(U/V) \geq L$ if and only if $C_G(U/V) \geq C_p$. By the FC-solidity condition this means that U/V is $\mathfrak{F}$-central if and only if $L \leq C_G(U/V)$.

Now suppose that X/Y is an $\mathfrak{F}$-central p-chief factor of G with $X \leq A$. Writing $\bar{A}_p = A_p/Y$ for the Sylow p-subgroup of A/Y and considering the action of the p′-group L/A on each finite p-subgroup F of A/Y and then applying Fitting's Lemma ([55] ,p.56) we have $F = [F,L] \times C_F(L)$. It follows that $\bar{A}_p = [\bar{A}_p,L] \times C_{\bar{A}_p}(L)$. Since L centralizes U/V, we have $[\bar{A}_p,L] < \bar{A}_p$ and each chief factor of G between $[\bar{A}_p,L]$ and $\bar{A}_p$ is centralized by L and so is $\mathfrak{F}$-central. But then $A_{p'}[A_p,L] < A$ and $G/A_{p'}[A_p,L]$ is an $\mathfrak{F}$-group contrary to A being the $\mathfrak{F}$-residual of G.

(ii) Let D be an $\mathfrak{F}$-normalizer of G; then by Theorem 6.12, DA = G and D avoids each chief factor below A. Hence $D \cap A = 1$ and D is a complement to A.

Let K be any other complement to A in G. If $d \in D$, there is a unique $k = k(d) \in K$ such that $dk^{-1} \in A$. Define $\phi: G \to G$ by $(ad)\phi = ak(d)$. Then ϕ

is an automorphism of G, since

$$\begin{aligned}((a_1d_1)(a_2d_2))\phi &= (a_1d_1a_2d_1^{-1}.d_1d_2)\phi \\ &= a_1d_1a_2d_1^{-1}.k_1k_2 \\ &= a_1(k_1a_2k_1^{-1})k_1k_2 \\ &= a_1k_1a_2k_2 \\ &= (a_1d_1)\phi.(a_2d_2)\phi\ ,\end{aligned}$$

and ϕ is clearly bijective. Therefore $K = D\phi$ is also an $\mathfrak{F}$-normalizer of G.

(iii) It is sufficient to show that $H \cap A = 1$ so that H is a complement to A. If U is a subgroup of $H \cap A$ with $U \lhd H$ then $U \lhd AH = G$. Thus a p-chief factor U/V of H with $U \leq H \cap A$ is a p-chief factor of G. Since $C_G(U/V) = AC_H(U/V)$, we have $G/C_G(U/V) \cong H/C_H(U/V)$. Since H is an $\mathfrak{F}$-group, $H/C_H(U/V) \in \mathfrak{f}(p)$ contrary to G having no $\mathfrak{F}$-central chief factor below A. □

We will make most use of $\mathfrak{F}$-normalizers in considering covering properties of subgroups in the finite factor groups. This will require the $\mathfrak{F}$-normalizers of G to be related to the $\mathfrak{F}$-normalizers of the finite factor groups of G.

Theorem 6.15. *Let $\mathfrak{F}$ be a saturated formation of $\mathfrak{S}$-groups.*

(i) *If N_i, $i \in I$, are normal subgroups of the $\mathfrak{S}$-group G and D is an $\mathfrak{F}$-normalizer of G, then $\bigcap_{i\in I}(N_iD) = (\bigcap_{i\in I}N_i)D$.*

(ii) *If $\{N_i : i \in I\}$ is a residual system of normal subgroups of finite index in the residually finite $\mathfrak{S}$-group G and, for each $i \in I$, D_i/N_i is an $\mathfrak{F}$-normalizer of G/N_i such that $D_i \geq D_j$ whenever $N_i \geq N_j$, then $\bigcap_{i\in I}D_i$ is an $\mathfrak{F}$-normalizer of G.*

Proof. (i) Using Theorem 6.12 (ii) it is clearly sufficient to consider the case in which $\bigcap_{i\in I}N_i = 1$. With the usual notation, let $D = S_\pi \cap \bigcap_{p\in\pi} N_G(C_p \cap S_{p'})$; then

$$\begin{aligned}\bigcap_{i\in I}(N_iD) &= \bigcap_{i\in I}N_i(S_\pi \cap \bigcap_{p\in\pi}N_G(C_p \cap S_{p'})) \\ &\leq \bigcap_{i\in I}N_iS_\pi \cap \bigcap_{p\in\pi} \bigcap_{i\in I}N_iN_G(C_p \cap S_{p'}) \\ &= S_\pi \cap \bigcap_{p\in\pi}N_G(C_p \cap S_{p'}), \text{ using Lemmas 5.5 and 5.13,} \\ &= D.\end{aligned}$$

The reverse inequality is clearly true, and so we have $\bigcap_{i\in I}(N_iD) = D$.

(ii) Let D be an $\mathfrak{F}$-normalizer of D; then for each $i \in I$, DN_i/N_i is an $\mathfrak{F}$-normalizer of G/N_i and so is conjugate to D_i/N_i. For each $i \in I$, let A_i be the set of inner automorphisms of G/N_i which map D_i/N_i onto DN_i/N_i. If $N_i \leqslant N_j$ and $\alpha \in A_i$ then the inner automorphism of G/N_j induced by α maps D_iN_j/N_j onto DN_j/N_j. But D_iN_j/N_j is an $\mathfrak{F}$-normalizer of G/N_j contained in D_j/N_j and so $D_iN_j = D_j$ and the automorphism induced by α is in A_j. By Theorem 4.17, there is a locally inner automorphism ϕ of G such that $D_i\phi = DN_i$, for each $i \in I$. Hence, by part (i), $(\bigcap_{i\in I}D_i)\phi = \bigcap_{i\in I}D_i\phi = \bigcap_{i\in I}DN_i = D$. Thus $\bigcap_{i\in I}D_i$ is locally conjugate to the $\mathfrak{F}$-normalizer D and so is itself an $\mathfrak{F}$-normalizer of G. □

If $\mathfrak{X}$ is any class of groups an $\mathfrak{X}$-*projector* of a group G is an $\mathfrak{X}$-subgroup X of G such that, whenever $X \leqslant H \leqslant G$ and $H/K \in \mathfrak{X}$, X covers H/K; that is, XK = H.

The following properties of $\mathfrak{X}$-projectors are immediate.

<u>Lemma 6.16</u>. <u>Let X be an $\mathfrak{X}$-projector of the group G. Then</u>

(i) X <u>is a maximal $\mathfrak{X}$-subgroup of</u> G,

(ii) <u>if</u> $X \leqslant H \leqslant G$, <u>then</u> X <u>is an $\mathfrak{X}$-projector of</u> H,

(iii) <u>if $\mathfrak{X}$ is Q-closed and</u> $N \triangleleft G$, <u>then</u> XN/N <u>is an $\mathfrak{X}$-projector of</u> G/N. □

As in the finite case, we make use of the following lemma.

<u>Lemma 6.17</u>. (Gaschütz [38]) <u>Let $\mathfrak{X}$ be a Q-closed class of groups and</u> N <u>a normal subgroup of the group</u> G. <u>If</u> $\bar{X}/N$ <u>is an $\mathfrak{X}$-projector of</u> G/N <u>and</u> X <u>is an $\mathfrak{X}$-projector of</u> $\bar{X}$, <u>then</u> X <u>is an $\mathfrak{X}$-projector of</u> G.

<u>Proof</u>. Since $\bar{X}/N$ is an $\mathfrak{X}$-group and $X \leqslant \bar{X}$, we have $XN = \bar{X}$. Now suppose that $X \leqslant H \leqslant G$ and $H/K \in \mathfrak{X}$. Then $\bar{X} = XN \leqslant HN$ and

$$HN/KN \cong H/(H \cap KN) \in Q\,\mathfrak{X} = \mathfrak{X}.$$

Thus $\bar{X}K = HN$ and so $H = (\bar{X} \cap H)K$. Also

$$(\bar{X} \cap H)/(\bar{X} \cap K) \cong K(\bar{X} \cap H)/K = H/K \in \mathfrak{X}$$

and $\bar{X} \cap H \geq X$. Therefore $\bar{X} \cap H = X(\bar{X} \cap K)$ and so $H = (\bar{X} \cap H)K = XK$, as required. □

Our aim is to show that if $\mathcal{F}$ is a saturated formation of $\mathfrak{S}$-groups then any $\mathfrak{S}$-group G has $\mathcal{F}$-projectors and these form a local conjugacy class so that each $\mathfrak{S}$-group has two (usually different) local conjugacy classes of $\mathcal{F}$ -subgroups, the $\mathcal{F}$-normalizers and the $\mathcal{F}$-projectors. We first show that soluble $\mathfrak{S}$-groups have $\mathcal{F}$-projectors and hence the finite factor groups of G have $\mathcal{F}$-projectors and then construct $\mathcal{F}$-projectors of G from these by using an inverse limit argument.

Lemma 6.18. Let $\mathcal{F}$ be a saturated formation of $\mathfrak{S}$-groups and let the $\mathfrak{S}$-group G have an abelian $\mathcal{F}$-residual A. Then the $\mathcal{F}$-normalizers of G are $\mathcal{F}$-projectors and, conversely, an $\mathcal{F}$-projector of G is an $\mathcal{F}$-normalizer.

Hence the $\mathcal{F}$-projectors of G are locally conjugate in G.

Proof. Let D be an $\mathcal{F}$-normalizer of G and suppose that $D \leq H \leq G$ and $H/K \in \mathcal{F}$. Now $HA \geq DA = G$, by Theorem 6.12, and so $H/(H \cap A) \cong G/A \in \mathcal{F}$. Therefore $H/(A \cap K) = H/(K \cap (H \cap A)) \in \mathcal{F}$. Also $A \cap K \lhd AH = G$ and so $D(A \cap K)/(A \cap K)$ is an $\mathcal{F}$-normalizer of $G/(A \cap K)$ contained in $H/(A \cap K)$. It follows from Theorem 6.14 (iii) that $D(A \cap K) = H$ so that $DK = H$ and D is an $\mathcal{F}$-projector of G.

Conversely, let F be an $\mathcal{F}$-projector of G; then $FA = G$ and so, by Theorem 6.14 (iii), F is an $\mathcal{F}$-normalizer of G. □

Theorem 6.19. Let $\mathcal{F}$ be a saturated formation of $\mathfrak{S}$-groups and let G be a soluble $\mathfrak{S}$-group. Then G has $\mathcal{F}$-projectors and any two $\mathcal{F}$-projectors of G are locally conjugate in G.

Proof. By induction on the length of an abelian normal series of G we may assume that G has an abelian normal subgroup A such that G/A has locally conjugate $\mathcal{F}$-projectors. Let $\bar{F}/A$ be an $\mathcal{F}$-projector of G/A; then $\bar{F}$ has an abelian $\mathcal{F}$-residual and so, by Lemma 6.18, $\bar{F}$ has an $\mathcal{F}$-projector F, say. By Lemma 6.17, F is an $\mathcal{F}$-projector of G.

Now let F_1 be a second $\mathcal{F}$-projector of G. By Lemma 6.16, FA/A and F_1A/A are $\mathcal{F}$-projectors of G/A and so, by the induction hypothesis, are locally conjugate in G/A. By Theorem 4.18(ii), there is a locally inner automorphism ϕ of G such that $(F_1A)\phi = FA$ and so $F_1\phi \leq FA$. By Lemma 6.16, F and $F_1\phi$ are $\mathfrak{F}$-projectors of FA and therefore, by Lemma 6.18, F and $F_1\phi$ are locally conjugate in FA and hence, using Theorem 4.18(i), F and $F_1\phi$ are locally conjugate in G. Therefore F and F_1 are locally conjugate in G. □

When we construct an $\mathfrak{F}$-projector F of a residually finite $\mathfrak{S}$-group G from the $\mathfrak{F}$-projectors of the finite factor groups of G our main problem will be to show that F has the necessary covering property. We do this by showing that, for any $H \geq F$, F contains an $\mathfrak{F}$-normalizer of H. First we need to consider $\mathfrak{F}$-abnormal subgroups.

Every maximal subgroup M of an $\mathfrak{S}$-group G has finite prime power index. If $|G:M| = p^{\alpha}$, we say that M is a *p-maximal* subgroup of G. In this case, G/M_G is a finite soluble group with a unique minimal normal subgroup N/M_G which is self-centralizing and is a normal complement to M/M_G([55],p.159). The following is an immediate consequence of this remark.

Lemma 6.20. Let $\mathfrak{f}$ be a formation function on the set π and let $\mathfrak{F} = \mathfrak{F}(\mathfrak{f})$. Let M be a p-maximal subgroup of the $\mathfrak{S}$-group G and let N/M_G be the unique minimal normal subgroup of G/M_G. Then the following are equivalent:

(a) $G/M_G \in \mathfrak{F}$,

(b) $p \in \pi$ and $G/N \in \mathfrak{f}(p)$,

(c) $p \in \pi$ and $M/M_G \in \mathfrak{f}(p)$,

(d) N/M_G is an $\mathfrak{F}$-central chief factor of G. □

If the p-maximal subgroup M satisfies these conditions then M is said to be an *$\mathfrak{F}$-normal maximal subgroup*; otherwise M is called an *$\mathfrak{F}$-abnormal maximal subgroup*. In particular, if $p \notin \pi$ then a p-maximal subgroup is $\mathfrak{F}$-abnormal. A normal p-maximal subgroup is $\mathfrak{F}$-normal if and only if $p \in \pi$. The definition is extended to non-maximal subgroups of G by defining a subgroup A of G to be *$\mathfrak{F}$-abnormal* in G if, whenever $A \leq V < U \leq G$ and V is maximal in U, V is an $\mathfrak{F}$-abnormal maximal subgroup of U. It is immediate that every subgroup B containing A is also an $\mathfrak{F}$-abnormal subgroup.

If $\mathfrak{F}$ is the class of locally nilpotent periodic FC-groups, then a maximal subgroup is $\mathfrak{F}$-normal if and only if it is normal in G. Thus the definition of an $\mathfrak{F}$-abnormal subgroup A coincides with that of a quasiabnormal subgroup (that is, each subgroup containing A is self-normalizing) and for $\mathfrak{S}$-groups this is equivalent to the usual definition of an abnormal subgroup.

Lemma 6.21. (Stonehewer [93]) Let A be a subgroup of the $\mathfrak{S}$-group G. Then A is abnormal in G if and only if it is quasiabnormal in G.

Proof. An abnormal subgroup is always quasiabnormal; so suppose that A is quasiabnormal in G and let $x \in G$. We must show that $x \in \langle A, x^{-1}Ax \rangle$. There is a finite normal subgroup F of G containing x and $\langle A, x^{-1}Ax \rangle \leq AF$. Let B be the core of A in AF. Clearly $|AF:A|$ is finite and so AF/B is a finite soluble group. Also A is quasiabnormal in AF and hence A/B is quasiabnormal in AF/B. By the result for finite soluble groups ([55], p.733), $xB \in \langle A, x^{-1}Ax \rangle /B$ and so $x \in \langle A, x^{-1}Ax \rangle$. □

Since locally nilpotent FC-groups are hypercentral, they have no proper self-normalizing subgroups and hence no proper abnormal subgroups. This fact can be generalized for any saturated formation $\mathfrak{F}$.

Theorem 6.22. Let $\mathfrak{F}$ be a saturated formation of $\mathfrak{S}$-groups. Then an $\mathfrak{F}$-group G has no proper $\mathfrak{F}$-abnormal subgroups.

Proof. Suppose that A is an $\mathfrak{F}$-abnormal subgroup of G. Then there is a subgroup B of G such that A is an $\mathfrak{F}$-abnormal maximal subgroup of B. Let $\{N_i/Z : i \in I\}$ be a residual system of normal subgroups of finite index in G/Z. If $BN_i = G$, for all i, then $B/(B \cap N_i) \cong G/N_i \in Q\mathfrak{F} = \mathfrak{F}$ and so $B/(B \cap Z) \in R\mathfrak{F} \cap \mathfrak{S} = \mathfrak{F}$. Since Z is a π-group it follows from Lemma 6.5 that $B \in \mathfrak{F}$. But it is clear from the definition that every maximal subgroup of an $\mathfrak{F}$-group is $\mathfrak{F}$-normal. This contradiction shows that, for some $i \in I$, $BN_i < G$ and hence there is an $\mathfrak{F}$-abnormal maximal subgroup of G containing BN_i. This is again contrary to G being an $\mathfrak{F}$-group. □

Theorem 6.23. Let $\mathfrak{F}$ be a saturated formation of $\mathfrak{S}$-groups and let F be an $\mathfrak{F}$-subgroup of the $\mathfrak{S}$-group G. Then F is an $\mathfrak{F}$-projector of G if and only if F is $\mathfrak{F}$-abnormal in G.

Proof. Let F be an $\mathfrak{F}$-projector of G and let U,V be subgroups of G such that $F \leqslant V < U \leqslant G$ and V is a p-maximal subgroup of U. If V is an $\mathfrak{F}$-normal maximal subgroup of U, then $p \in \pi$ and $U/K \in \mathfrak{F}$, where K is the core of V in U. But then, by the covering property of F, we have $FK = U$ contrary to $FK \leqslant V < U$.

Conversely, let F be an $\mathfrak{F}$-abnormal $\mathfrak{F}$-subgroup of G and let H,K be subgroups of G such that $F \leqslant H$, $K \triangleleft H$ and $H/K \in \mathfrak{F}$. If $FK < H$ then FK is $\mathfrak{F}$-abnormal in G and hence in H. Therefore FK/K is a proper $\mathfrak{F}$-abnormal subgroup of H/K contrary to Theorem 6.22. □

It is therefore necessary for us to show that an $\mathfrak{F}$-abnormal subgroup of an $\mathfrak{S}$-group G contains an $\mathfrak{F}$-normalizer of G.

Lemma 6.24. *Let $\mathfrak{F}$ be the saturated formation of $\mathfrak{S}$-groups defined by the formation function $\mathfrak{f}$ on the set π and, for some $p \in \pi$, let C_p be the $(\mathfrak{f}(p),p)$-centralizer of the $\mathfrak{S}$-group G. If M is an $\mathfrak{F}$-abnormal maximal subgroup of G containing the Sylow p'-subgroup $S_{p'}$, then $M \geqslant N_G(C_p \cap S_{p'})$.*

Proof. Let $K = M_G$ and N/K the unique minimal normal subgroup of the finite soluble group G/K. Since M is $\mathfrak{F}$-abnormal in G, $G/K \notin \mathfrak{F}$ and $G/N \notin \mathfrak{f}(p)$. Also $C_G(N/K) = N$ and so $N \not\leqslant C_p$. Thus $KC_p > K$ and so KC_p/K contains the unique minimal normal subgroup N/K of G/K. Hence $KC_p = NC_p$. It follows that

$$N(M \cap C_p) = NK(M \cap C_p) = N(M \cap KC_p) = N(M \cap NC_p) = NM \cap NC_p = NC_p.$$

Now NC_p/N contains a minimal normal subgroup Q/N of G/N and, since $N = C_G(N/K)$, Q/N must be a q-group for some $q \neq p$. Thus $(Q \cap M)/K$ is a normal p'-subgroup of M/K and so is contained in each Sylow p'-subgroup of M/K; in particular $Q \cap M \leqslant KS_{p'}$ and hence $Q \cap M = Q \cap KS_{p'}$.
We now have

$$\begin{aligned} N_G(C_p \cap S_{p'}) &\leqslant N_G(KC_p \cap KS_{p'}), \\ &\leqslant N_G(KC_p \cap KS_{p'} \cap Q), \\ &= N_G(KC_p \cap M \cap Q), \\ &= N_G(K(C_p \cap M) \cap Q), \end{aligned}$$

$$= N_G((N \cap M)(C_p \cap M) \cap Q),$$
$$= N_G(N(C_p \cap M) \cap M \cap Q),$$
$$= N_G(NC_p \cap M \cap Q),$$
$$= N_G(M \cap Q).$$

But $M \cap Q \lhd M$ and since $M_G < M \cap Q$, we have $N_G(M \cap Q) = M$, as required. □

Lemma 6.25. Let $\mathfrak{F}$ be the saturated formation of $\mathfrak{S}$-groups defined by the formation function $\mathfrak{f}$ on the set π and, for each $p \in \pi$, write $C_p(X)$ for the $(\mathfrak{f}(p),p)$-centralizer of an $\mathfrak{S}$-group X. If M is an $\mathfrak{F}$-abnormal maximal subgroup of the $\mathfrak{S}$-group G, then $C_p(M) \leq C_p(G) \cap M$, for each $p \in \pi$.

Proof. Let U/V be any p-chief factor of G with $G/C_G(U/V) \in \mathfrak{f}(p)$. We must show that $C_p(M)$ centralizes U/V. If M is a q-maximal subgroup and $q \neq p$ then it is clear that M covers U/V. If M is a p-maximal subgroup then, by Lemma 6.24, there is a Sylow p′-subgroup $S_{p'}$ of G such that $M \geq N_G(C_p \cap S_{p'})$ and hence, by Lemma 6.10, M covers U/V. In both cases we have $U/V \overset{M}{\cong} (U \cap M)/(V \cap M)$. Now $G/C_G(U/V) \in \mathfrak{f}(p) \subseteq \mathfrak{F}$ and if $C_G(U/V) \leq M$ we would have $C_G(U/V) \leq M_G$ and hence $G/M_G \in \mathfrak{F}$ contrary to M being $\mathfrak{F}$-abnormal. Therefore $MC_G(U/V) = G$ and $M/C_M(U/V) \cong G/C_G(U/V) \in \mathfrak{f}(p)$. It follows that every p-chief factor of M between $V \cap M$ and $U \cap M$ is $\mathfrak{F}$-central in M and so is centralized by $C_p(M)$. If $C_p(M)$ does not centralize U/V then $C_p(M)\, C_M(U/V)/C_M(U/V)$ is non-trivial and so contains a non-trivial q-subgroup $Q/C_M(U/V)$ of $M/C_M(U/V)$. Since $O_p(M/C_M(U/V)) \cong O_p(G/C_G(U/V))$ is trivial, we have $q \neq p$. Now $Q/C_M(U/V)$ is a finite q-group acting on the finite p-group $(U \cap M)/(V \cap M)$ and centralizing each chief factor of M between $V \cap M$ and $U \cap M$. But applying Fitting's Lemma ([55] , p.350) this implies that Q centralizes $(U \cap M)/(V \cap M)$. This is contrary to $Q > C_M(U/V)$ and so $C_p(M)$ must centralize U/V. □

Theorem 6.26. Let $\mathfrak{F}$ be a saturated formation of $\mathfrak{S}$-groups and let A be an $\mathfrak{F}$-abnormal subgroup of the $\mathfrak{S}$-group G. Let D_o be the $\mathfrak{F}$-normalizer of A associated with the Sylow basis $\mathcal{T}$ of A. Then there is a Sylow basis $\mathcal{S}$ of G such that $\mathcal{S} \cap A = \mathcal{T}$ and if D is the $\mathfrak{F}$-normalizer of G associated with $\mathcal{S}$, then $D \leq D_o$.

Proof. By Lemma 5.24, there is a Sylow basis S of G such that $S \cap A = T$. We have only to prove that $D \leq D_o$.

We suppose first that A has finite index in G. A simple induction argument on the index $|G:A|$ reduces this to the case in which A is an $\mathcal{F}$-abnormal maximal subgroup of G. In this case, Lemma 6.24 shows that $D \leq A$ and so, with the usual notation,

$$\begin{aligned} D &= S_\pi \cap A \cap \bigcap_{p\in\pi} N_A(C_p \cap S_{p'}), \\ &\leq S_\pi \cap A \cap \bigcap_{p\in\pi} N_A(A \cap C_p \cap S_{p'}), \\ &\leq S_\pi \cap A \cap \bigcap_{p\in\pi} N_A(C_p(A) \cap S_{p'}), \text{ by Lemma 6.25}, \\ &= D_o. \end{aligned}$$

We now consider the general case. Let $\{N_i/Z : i \in I\}$ be a residual system of normal subgroups of finite index in G/Z. Then, for each $i \in I$, DN_i/N_i is the $\mathcal{F}$-normalizer of G/N_i associated with SN_i/N_i, D_oN_i/N_i is the $\mathcal{F}$-normalizer of AN_i/N_i associated with TN_i/N_i and AN_i/N_i is an $\mathcal{F}$-abnormal subgroup of the finite soluble group G/N_i. By the case considered above, $DN_i \leq D_oN_i$ and hence, using Theorem 6.15, $DZ = \bigcap_{i\in I} DN_i \leq \bigcap_{i\in I} D_oN_i = D_oZ$. But $A \geq Z$ and so $D_o = S_\pi \cap D_oZ$ and $D = S_\pi \cap DZ$. Hence $D \leq D_o$, as required. □

We can now prove the main result of the formation theory that we have developed, namely that every $\mathfrak{S}$-group does have a local conjugacy class of $\mathcal{F}$-projectors.

Theorem 6.27. (Tomkinson [99]) Let $\mathcal{F}$ be a saturated formation of $\mathfrak{S}$-groups. Then

(i) an $\mathfrak{S}$-group G has $\mathcal{F}$-projectors,

(ii) if F is an $\mathcal{F}$-projector of G and $N_i, i \in I$, are normal subgroups of G, then $\bigcap_{i\in I} FN_i = F(\bigcap_{i\in I} N_i)$,

(iii) any two $\mathcal{F}$-projectors of G are locally conjugate in G.

Proof. (i) Let $\{N_i/Z : i \in I\}$ be a residual system of normal subgroups of finite index in G/Z. By Theorem 6.19, each finite soluble group G/N_i has

$\mathcal{F}$-projectors. For each $i \in I$, let X_i be the finite non-empty set consisting of all $\mathcal{F}$-projectors of G/N_i. The index set I can be made into a directed set by defining $i \prec j$ to mean $N_i \geq N_j$. If $i \prec j$ and $F/N_j \in X_j$, then $FN_i/N_i \in X_i$, by Lemma 6.16 (iii), and we can define $\theta_{ji}:X_j \to X_i$ by $(F/N_j)\theta_{ji} = FN_i/N_i$. These definitions make $\{X_i : i \in I\}$ into an inverse system of finite non-empty sets. By Theorem 5.21, the inverse limit is non-empty and so there is a set $\{F_i/N_i : i \in I\}$ such that each F_i/N_i is an $\mathcal{F}$-projector of G/N_i and, whenever $N_i \geq N_j, F_jN_i = F_i$. We define $F = \bigcap_{i \in I} F_i$ and prove that F/Z is an $\mathcal{F}$-projector of G/Z. Most of the required properties will be proved by showing that F/Z contains an $\mathcal{F}$-normalizer of an appropriate subgroup.

(I) $FN_i = F_i$, for each $i \in I$.

Fix $i \in I$ and let $J = \{j \in I : N_j \leq N_i\}$. For each $j \in J$, F_j/N_j is an $\mathcal{F}$-projector of G/N_j contained in F_i/N_j and so, by Lemma 6.16 (i), is an $\mathcal{F}$-projector of F_i/N_j. By Theorem 6.23 and 6.26, F_j/N_j contains an $\mathcal{F}$-normalizer of F_i/N_j. For each $j \in J$, let W_j be the finite non-empty set of $\mathcal{F}$-normalizers of F_i/N_j which are contained in F_j/N_j. The index set J can be made into a directed set by defining $j \prec k$ to mean $N_j \geq N_k$. If $j \prec k$ and $D_k/N_k \in W_k$, then D_kN_j/N_j is an $\mathcal{F}$-normalizer of F_i/N_j and $D_kN_j \leq F_kN_j = F_j$. Hence $D_kN_j/N_j \in W_j$ and we can define $\theta_{kj}:W_k \to W_j$ by $(D_k/N_k)\theta_{kj} = D_kN_j/N_j$. These definitions make $\{W_j : j \in J\}$ into an inverse system of finite non-empty sets. By Theorem 5.21, the inverse limit is non-empty and so there is a set $\{D_j/N_j : j \in J\}$ such that each D_j/N_j is an $\mathcal{F}$-normalizer of F_i/N_j contained in F_j/N_j and, whenever $N_j \geq N_k$, $D_kN_j = D_j$. By Theorem 6.15, $D/Z = \bigcap_{j \in J} D_j/Z$ is an $\mathcal{F}$-normalizer of F_i/Z and it is clear that $D \leq F$. Since $F_i/N_i \in \mathcal{F}$, Theorem 6.12 shows that $DN_i = F_i$ and hence $FN_i = F_i$.

(II) F/Z is an $\mathcal{F}$-group.

We have $F/(F \cap N_i) \cong FN_i/N_i = F_i/N_i \in \mathcal{F}$. Therefore $F/Z \in R\mathcal{F} \cap \mathcal{S} = \mathcal{F}$.

(III) If $Z \leq K \leq H$, $H \geq F$ and $H/K \in \mathcal{F}$, then $FK = H$.

This is proved by showing that F/Z contains an $\mathcal{F}$-normalizer of H/Z. For each $i \in I$, $HN_i \geq FN_i = F_i$ and so F_i/N_i is an $\mathcal{F}$-projector of HN_i/N_i. Under the isomorphism between HN_i/N_i and $H/(H \cap N_i)$, we see that $(F_i \cap H)/(N_i \cap H)$ is an $\mathcal{F}$-projector of $H/(N_i \cap H)$. For each $i \in I$, let V_i be the finite non-empty set consisting of the $\mathcal{F}$-normalizers of $H/(N_i \cap H)$ which are contained in $(F_i \cap H)/(N_i \cap H)$. As above, $\{V_i : i \in I\}$ can be made into an inverse system by defining $i \prec j$ to mean $N_i \geq N_j$ and, whenever $i \prec j$, defining $\theta_{ji} : V_j \to V_i$ by $(D_j/(N_j \cap H))\theta_{ji} = D_j(N_i \cap H)/(N_i \cap H)$. Since the inverse limit is non-empty, there is a set $\{D_i/(N_i \cap H) : i \in I\}$ such that each $D_i/(N_i \cap H)$ is an $\mathcal{F}$-normalizer of $H/(N_i \cap H)$ contained in $(F_i \cap H)/(N_i \cap H)$ and, whenever $N_j \leq N_i$, $D_j(N_i \cap H) = D_i$. By Theorem 6.15, $D/Z = \bigcap_{i \in I}(D_i/Z)$ is an $\mathcal{F}$-normalizer of H/Z and clearly $D \leq F$. Theorem 6.12 shows that $DK = H$ and hence $FK = H$, as required.

We have now shown that F/Z is an $\mathcal{F}$-projector of G/Z. It is clear that the unique Sylow π-subgroup of F is an $\mathcal{F}$-projector of F and hence, using Lemma 6.17, is also an $\mathcal{F}$-projector of G.

(ii) Let $F^* = \bigcap_{i \in I}(FN_i)$ and let $N = \bigcap_{i \in I} N_i$; then $F^* \geq FN$ and so $F^* N_i = FN_i$, for all $i \in I$. Therefore

$$F^*/(F^* \cap N_i) \cong F^* N_i/N_i = FN_i/N_i \cong F/(F \cap N_i) \in Q\mathcal{F} = \mathcal{F}$$

and so $F^*/N \in R\mathcal{F} \cap \mathfrak{S} = \mathcal{F}$. By the maximality of the $\mathcal{F}$-projector FN/N, we must have $F^* = FN$.

(iii) Let F_1 and F_2 be two $\mathcal{F}$-projectors of the $\mathfrak{S}$-group G. Then F_1 and F_2 are the unique Sylow π-subgroups of F_1Z and F_2Z, respectively. It is therefore sufficient to prove that F_1Z and F_2Z are locally conjugate in G. Let $\{N_i/Z : i \in I\}$ be a residual system of normal subgroups of finite index in G/Z. Then F_1N_i/N_i and F_2N_i/N_i are $\mathcal{F}$-projectors of the finite soluble group G/N_i and so, using Theorem 6.19, F_1N_i/N_i and F_2N_i/N_i are conjugate in G/N_i. For each $i \in I$, let A_i be the set of inner automorphisms of G/N_i which map F_1N_i/N_i onto F_2N_i/N_i. Then these sets satisfy the conditions of Theorem 4.17 and so there is a locally inner automorphism ϕ of G/Z such that $(F_1\phi)N_i/Z = F_2N_i/Z$ for each $i \in I$. Using Theorem 4.18, we see that there is a locally inner automorphism θ of G such that $(F_1\phi)N_i = F_2N_i$, for each $i \in I$, and hence $(F_1\theta)Z = \bigcap_{i \in I}(F_1\theta)N_i = \bigcap_{i \in I} F_2N_i = F_2Z$, as required. □

We have shown that every $\mathfrak{S}$-group has a unique local conjugacy class of $\mathcal{F}$-abnormal $\mathcal{F}$-subgroups. For the case in which $\mathcal{F}$ is the class of locally nilpotent groups we obtain the abnormal locally nilpotent subgroups. These subgroups can be given the following simpler characterization which is much more familiar in the finite case.

Theorem 6.28. A locally nilpotent subgroup of an $\mathfrak{S}$-group G is an $L\mathfrak{N}$-projector of G if and only if it is self-normalizing.

Proof. Certainly an $L\mathfrak{N}$-projector is abnormal and therefore self-normalizing. Conversely, let F be a self-normalizing locally nilpotent subgroup of the $\mathfrak{S}$-group G. If F is not abnormal in G, then there is a subgroup V containing F such that $V < N_G(V)$. Let U/V be a finite non-trivial nilpotent subgroup of $N_G(V)/V$ and let X be a finite normal subgroup of U such that U = VX. Then $XF/(V \cap XF) \cong U/V$ is nilpotent and F is a self-normalizing locally nilpotent subgroup of finite index in XF. If K is the core of F in XF, then F/K is a self-normalizing nilpotent subgroup of the finite soluble group XF/K. Therefore $F(V \cap XF) = XF$, contrary to $F \lneq V \cap XF < XF$ and so F must be abnormal in G. □

The self-normalizing nilpotent subgroups of a finite soluble group G are usually called the Carter subgroups of G. The intersection of Carter subgroups is the hypercentre of G. Also the Carter subgroups generate the whole of G and have a considerable influence on the structure of G. The most striking result of this type is due to Dade [18] and says that if the order of a Carter subgroup of a finite soluble group G is the product of n (not necessarily distinct) primes, then the Fitting length of G is at most f(n), where f is some function depending only on n.

It is tempting to believe that the Carter subgroups of an $\mathfrak{S}$-group G will have considerable influence on the structure of G. It is not even clear what type of structure is appropriate for consideration here but we note two almost trivial results which may be a pointer.

Theorem 6.29. Let F be a Carter subgroup of the $\mathfrak{S}$-group G.

(i) If $\mathfrak{m}$ is an infinite cardinal and $|F| < \mathfrak{m}$, then $|G| < \mathfrak{m}$.

(ii) If F has finite exponent, then G/Z(G) is embeddable in a direct product of finite groups.

Proof. We need only observe that FG′ = G. Then in (i), we have $|G/G'| < \mathfrak{m}$ so that Corollary 1.20 gives the required result. In (ii), we see that G/G′Z has finite exponent and the result follows from Corollary 2.29. □

One can suggest many different possible results of this type. We mention one possibility.

Question 6B. Let G be an $\mathfrak{S}$-group whose Carter subgroups are in the class *QSD* $\mathcal{F}$. Does it follows that G is in the class *QSD* $\mathcal{F}$?

Strong restrictions on G are obtained if we demand that the $\mathcal{F}$-projectors (or $\mathcal{F}$-normalizers) are conjugate in G rather than just locally conjugate. In order to simplify the statement of the theorem we restrict ourselves to the case in which G is a $\pi(\mathcal{F})$-group, where $\pi(\mathcal{F})$ is the set of primes on which the formation function defining $\mathcal{F}$ is defined.

Theorem 6.30. (Tomkinson [100]) Let $\mathcal{F}$ be a saturated formation of $\mathfrak{S}$-groups and let G be an $\mathfrak{S}$-group which is a $\pi(\mathcal{F})$-group. Then the following are equivalent:

(a) the $\mathcal{F}$-projectors of G are conjugate,

(b) the $\mathcal{F}$-normalizers of G are conjugate,

(c) $G \in \mathfrak{S}^{*}\mathcal{F}$,

(d) any one of conditions (a) - (c) hold in $G/G^{\mathfrak{A}\mathcal{F}}$, where $G^{\mathfrak{A}\mathcal{F}}$ is the (abelian-by-$\mathcal{F}$)-residual of G.

Proof. (a) or (b) implies (c). Since G/Z is a $\mathfrak{Y}$-group and the $\mathcal{F}$-projectors ($\mathcal{F}$-normalizers) of G are conjugate it follows from Theorem 4.26(ii) that G/Z has only finitely many $\mathcal{F}$-projectors ($\mathcal{F}$-normalizers). If X/Z is an $\mathcal{F}$-projector ($\mathcal{F}$-normalizer) of G/Z, then $|G:N_G(X)|$ is finite and so G has a finite normal subgroup F such that $FN_G(X) = G$. Since G is a π-group, X is an $\mathcal{F}$-projector ($\mathcal{F}$-normalizer) of G and so XF/F is an $\mathcal{F}$-projector ($\mathcal{F}$-normalizer) of G/F. But $XF \triangleleft FN_G(X) = G$ and so $XF = G$. Therefore $G/F \cong X/(X \cap F) \in Q\mathcal{F} = \mathcal{F}$.

(c) implies (a) and (b). If X is an $\mathcal{F}$-projector ($\mathcal{F}$-normalizer) of G, then $|G:X|$ is finite and so $C\ell(X)$ is finite. Hence $Lc\ell(X) = C\ell(X)$ and the $\mathcal{F}$-projectors ($\mathcal{F}$-normalizers) are conjugate.

(c) implies (d) is clear

(d) implies (c). Let M be the $\mathcal{F}$-residual of G so that M′ is the (abelian-by-$\mathcal{F}$)-residual of G. Then $G/M' \in \mathfrak{S}^{*}\mathcal{F}$ and so M/M′ is finite. By Corollary 1.20, M is finite and so $G \in \mathfrak{S}^{*}\mathcal{F}$. □

The special case of this result in which $\mathfrak{F}$ is the formation of locally nilpotent groups is of interest in that it gives conditions equivalent to the Sylow bases being conjugate.

Corollary 6.31. (Stonehewer [94]) The following conditions on the $\mathfrak{S}$-group G are equivalent:

(a) the Carter subgroups of G are conjugate,

(b) the basis normalizers of G are conjugate,

(c) G is finite-by-locally nilpotent,

(d) the Sylow bases of G are conjugate,

(e) G is (locally nilpotent)-by-finite.

Proof. The equivalence of (a), (b), (c) is proved in Theorem 6.30. It is immediate that (d) implies (b) and that (e) implies (d). It is therefore sufficient to note that (c) implies (e). Let G have a finite normal subgroup F such that G/F is locally nilpotent. Then G has a normal subgroup N of finite index such that $N \cap F \leq Z(G)$. Then $N/(N \cap F) \cong NF/F$ is locally nilpotent and since $N \cap F$ is central, N is therefore locally nilpotent. □

On the subject of the conjugacy of $\mathfrak{F}$-projectors it should be noted that $\mathfrak{F}$-projectors are L-pronormal and so Theorem 4.22 immediately gives us

Theorem 6.32. Let F_1 and F_2 be $\mathfrak{F}$-projectors of the $\mathfrak{S}$-group G. Then F_1 and F_2 are conjugate in G if and only if $|F_1:F_1 \cap F_2|$ is finite. □

There is one final characterization of $\mathfrak{F}$-projectors which should be given as this will strengthen the duality between $\mathfrak{F}$-projectors and the $\mathfrak{X}$-injectors which follow.

Theorem 6.33. Let $\mathfrak{F}$ be a saturated formation of $\mathfrak{S}$-groups. Then a subgroup F of the $\mathfrak{S}$-group G is an $\mathfrak{F}$-projector of G if and only if FN/N is a maximal $\mathfrak{F}$-subgroup of G/N, for each normal subgroup N of G.

Proof. If F is an $\mathfrak{F}$-projector of G, then FN/N is an $\mathfrak{F}$-projector of G/N and so is a maximal $\mathfrak{F}$-subgroup of G/N.

Conversely, suppose FN/N is a maximal $\mathfrak{F}$-subgroup of G/N, for each $N \triangleleft G$.

Let $\{N_i/Z : i \in I\}$ be a residual system of normal subgroups of finite index in G and, for each $i \in I$, let A_i/N_i be a minimal normal subgroup of G/N_i. Now FN_i/N_i and FA_i/A_i satisfy the conditions for F and, by induction, we may assume that FA_i/A_i is an $\mathcal{F}$-projector of G/A_i. Then FN_i/N_i is a maximal $\mathcal{F}$-subgroup of the (abelian-by-$\mathcal{F}$)-group FA_i/N_i. It follows from Theorem 6.14 (iii) that FN_i/N_i is an $\mathcal{F}$-normalizer of FA_i/N_i and so, by Lemma 6.18, FN_i/N_i is a $\mathcal{F}$-projector of FA_i/N_i and hence of G/N_i, by Lemma 6.17. It now follows that $FZ/Z = \bigcap_{i \in I} FN_i/Z$ is an $\mathcal{F}$-projector of G/Z. Since F is a maximal $\mathcal{F}$-subgroup of FZ, F must be the unique Sylow π-subgroup of FZ and so is an $\mathcal{F}$-projector of FZ. Using Lemma 6.17 again, we see that F is an $\mathcal{F}$-projector of G. □

The concept of a formation and of $\mathcal{F}$-projectors was dualized in the class of finite soluble groups by Fischer, Gaschütz and Hartley [33] . In this theory one is concerned with properties that are preserved in passing to normal subgroups and the use of an inverse limit argument using the finite normal subgroups of G is very much easier than the arguments used for the formation theory.

A class $\mathfrak{E}$ of finite soluble groups is called a *Fitting class* if

(FC*1) $\mathfrak{E}$ is S_n-closed; i.e., if $G \in \mathfrak{E}$ and $N \lhd G$, then $N \in \mathfrak{E}$; and

(FC*2) $\mathfrak{E}$ is N_o-closed; i.e., if $G = H_1H_2$ where H_1 and H_2 are normal $\mathfrak{E}$-subgroups of G, then $G \in \mathfrak{E}$.

If $\mathfrak{X}$ is any class of groups, then an $\mathfrak{X}$-*injector* of G is defined to be an $\mathfrak{X}$-subgroup X of G such that, for each subnormal subgroup S of G, $X \cap S$ is a maximal $\mathfrak{X}$-subgroup of S.

Theorem 6.34. (Fischer, Gaschütz and Hartley [33]) Let $\mathfrak{E}$ be a Fitting class of finite soluble groups. Then a finite soluble group G has $\mathfrak{E}$-injectors and any two $\mathfrak{E}$-injectors of G are conjugate in G. □

We show how this result can be extended to the class $\mathfrak{S}$ of locally soluble periodic FC-groups. Define a subclass $\mathfrak{X}$ of $\mathfrak{S}$ to be a *Fitting class* of $\mathfrak{S}$-groups if

(FC1) $\mathfrak{X}$ is S_n-closed; and

(FC2) $N\mathfrak{X} \cap \mathfrak{S} = \mathfrak{X}$; i.e., if G is an $\mathfrak{S}$-group which is the join of normal $\mathfrak{X}$-subgroups then G is an $\mathfrak{X}$-group.

These conditions ensure that G has a unique maximal normal $\mathfrak{X}$-subgroup $G_{\mathfrak{X}}$ called the $\mathfrak{X}$-radical of G. A subnormal subgroup S of G is an $\mathfrak{X}$-group if and only if $S \leq G_{\mathfrak{X}}$ ([84] Part 1, p.19).

The Fitting classes of $\mathfrak{S}$-groups can be determined completely in terms of Fitting classes of $\mathfrak{S}^*$-groups.

Theorem 6.35. (i) If $\mathfrak{E}$ is a Fitting class of $\mathfrak{S}^*$-groups, then $L\mathfrak{E} \cap \mathfrak{S}$ is a Fitting class of $\mathfrak{S}$-groups.

(ii) If $\mathfrak{X}$ is a Fitting class of $\mathfrak{S}$-groups, then $\mathfrak{X}^*$ is a Fitting class of $\mathfrak{S}^*$-groups and $\mathfrak{X} = L\mathfrak{X}^* \cap \mathfrak{S}$.

Proof. (i) Let N be a normal subgroup of the $(L\mathfrak{E} \cap \mathfrak{S})$-group G. If F is a finite subgroup of N, then F is contained in an $\mathfrak{E}$-subgroup E of G. But then $E \cap N \in S_n \mathfrak{E} = \mathfrak{E}$ and so F is contained in an $\mathfrak{E}$-subgroup of N. Thus $N \in L\mathfrak{E} \cap \mathfrak{S}$ and $L\mathfrak{E} \cap \mathfrak{S}$ is S_n-closed.

Let G be an $\mathfrak{S}$-group generated by the normal $(L\mathfrak{E} \cap \mathfrak{S})$-groups $H_i, i \in I$. If F is a finite subgroup of G, then F is contained in a finite product $H_{i(1)} \dots H_{i(r)}$, say, and we can choose finite normal subgroups X_j of G such that $X_j \leq H_{i(j)}$ and $F \leq X_1 \dots X_r$. But $X_j \in S_n(L\mathfrak{E} \cap \mathfrak{S}) \cap \mathfrak{S}^* = \mathfrak{E}$ and so $X_1 \dots X_r \in N_o \mathfrak{E} = \mathfrak{E}$. Thus F is contained in an $\mathfrak{E}$-subgroup of G and so $G \in L\mathfrak{E} \cap \mathfrak{S}$.

(ii) It is clear from the definitions that $\mathfrak{X}^*$ is a Fitting class of $\mathfrak{S}^*$-groups. If $G \in \mathfrak{X}$, then every finite normal subgroup of G is in $\mathfrak{X}^*$ and so $G \in L\mathfrak{X}^* \cap \mathfrak{S}$. Conversely, if $G \in L\mathfrak{X}^* \cap \mathfrak{S}$ then again every finite normal subgroup of G is contained in an $\mathfrak{X}^*$-group and so is itself an $\mathfrak{X}^*$-group. Therefore G is the join of normal $\mathfrak{X}$-groups and so is an $\mathfrak{X}$-group. □

Our use of an inverse limit argument applied to the finite normal subgroups of G depends on the $\mathfrak{X}$-injectors behaving well with respect to normal subgroups.

Lemma 6.36. Let $\mathfrak{X}$ be a Fitting class of $\mathfrak{S}$-groups and let $\{F_i : i \in I\}$ be a local system of finite normal subgroups of the $\mathfrak{S}$-group G. Then a subgroup V of G is an $\mathfrak{X}$-injector of G if and only if, for each $i \in I$, $V \cap F_i$ is an $\mathfrak{X}$-injector of F_i.

Proof. Let V be an $\mathfrak{X}$-injector of G; then $V \cap F_i \in S_n \mathfrak{X} = \mathfrak{X}$. If S is a subnormal subgroup of F_i, then S is subnormal in G and so $V \cap S$ is a maximal $\mathfrak{X}$-subgroup of S. Therefore $V \cap F_i$ is an $\mathfrak{X}$-injector of F_i.

Conversely, suppose that $V \cap F_i$ is an $\mathfrak{X}$-injector of F_i, for each $i \in I$. Then $V \cap F_i \triangleleft V$ and so $V = \bigcup_{i\in I}(V \cap F_i) \in N\mathfrak{X} \cap \mathfrak{S} = \mathfrak{X}$. Now let S be a subnormal subgroup of G and let W be an $\mathfrak{X}$-subgroup of S containing $V \cap S$. For each $i \in I$, $W \cap F_i$ is an $\mathfrak{X}$-subgroup of $S \cap F_i$ containing $V \cap S \cap F_i$. But since $V \cap F_i$ is an $\mathfrak{X}$-injector of F_i, $V \cap S \cap F_i$ is a maximal $\mathfrak{X}$-subgroup of $S \cap F_i$ and so $W \cap F_i = V \cap S \cap F_i$. Hence $W = V \cap S$, as required. □

Theorem 6.37. (Tomkinson [102]) Let $\mathfrak{X}$ be a Fitting class of $\mathfrak{S}$-groups and let G be an $\mathfrak{S}$-group. Then

(i) G has $\mathfrak{X}$-injectors,

(ii) any two $\mathfrak{X}$-injectors of G are locally conjugate in G,

(iii) if V is an $\mathfrak{X}$-injector of G and S is a serial subgroup of G then $V \cap S$ is an $\mathfrak{X}$-injector of G.

Proof. (i) Let $\{F_i : i \in I\}$ be a local system of finite normal subgroups of G. By Theorem 6.34, each F_i has $\mathfrak{X}$-injectors and so, for each $i \in I$, the set X_i of $\mathfrak{X}$-injectors of F_i is finite and non-empty. The set I can be made into a directed set by defining $i \prec j$ to mean $F_i \leqslant F_j$. If $i \prec j$ and V is an $\mathfrak{X}$-injector of F_j then $V \cap F_i$ is an $\mathfrak{X}$-injector of F_i and we can define $\theta_{ji} : X_j \to X_i$ by $V\theta_{ji} = V \cap F_i$. With these definitions, $\{X_i : i \in I\}$ becomes an inverse system of finite non-empty sets. The inverse limit is non-empty and so there is a set $\{V_i : i \in I\}$ such that each V_i is an $\mathfrak{X}$-injector of F_i and, whenever $F_i \leqslant F_j$, $V_j \cap F_i = V_i$.

We define $V = \bigcup_{i\in I} V_i$. Then clearly $V \cap F_i = V_i$ and, by Lemma 6.36, V is an $\mathfrak{X}$-injector of G.

(ii) Let V_1 and V_2 be two $\mathfrak{X}$-injectors of G. Then, for each $i \in I$, $V_1 \cap F_i$ and $V_2 \cap F_i$ are $\mathfrak{X}$-injectors of F_i and so are conjugate in F_i. Let A_i be the set of automorphisms of F_i which are induced by inner automorphisms of G and which map $V_1 \cap F_i$ onto $V_2 \cap F_i$. Then the sets A_i satisfy the conditions of Theorem 4.16 and so there is a locally inner automorphism ϕ of G such that $(V_1 \cap F_i)\phi = V_2 \cap F_i$, for all $i \in I$, and hence $V_1\phi = \bigcup_{i\in I}(V_1 \cap F_i)\phi =$
$= \bigcup_{i\in I}(V_2 \cap F_i) = V_2$.

(iii) For each $i \in I$, $S \cap F_i$ is subnormal in F_i and so $V \cap S \cap F_i$ is an

$\mathfrak{X}$-injector of $S \cap F_i$. Since $\{S \cap F_i : i \in I\}$ is a local system of finite normal subgroups of S it follows from Lemma 6.36 that $V \cap S$ is an $\mathfrak{X}$-injector of S. □

It was proved in [33] that if V is an $\mathfrak{E}$-injector of the finite soluble group G and H is a subgroup of G containing V, then V is an $\mathfrak{E}$-injector of H. This result extends easily to $\mathfrak{S}$-groups and leads to the pronormality conditions which were used in proving conjugacy results in Chapter 4.

Theorem 6.38. Let $\mathfrak{X}$ be a Fitting class of $\mathfrak{S}$-groups and let V be an $\mathfrak{X}$-injector of the $\mathfrak{S}$-group G. Then

(i) if $V \leq H \leq G$, then V is an $\mathfrak{X}$-injector of H,

(ii) V is L-pronormal in G,

(iii) let V_1 be a second $\mathfrak{X}$-injector of G; then V and V_1 are conjugate in G if and only if $|V:V \cap V_1|$ is finite.

Proof. (i) Let $\{F_i : i \in I\}$ be a local system of finite normal subgroups of G. Then $V \cap F_i$ is an $\mathfrak{X}$-injector of the finite soluble group F_i and hence is an $\mathfrak{X}$-injector of $H \cap F_i$. Applying Lemma 6.36 to the local system $\{H \cap F_i : i \in I\}$ of H we see that V is an $\mathfrak{X}$-injector of H.

(ii) Let ϕ be a locally inner automorphism of G. By part (i), V and $V\phi$ are $\mathfrak{X}$-injectors of $\langle V, V\phi \rangle$ and so are locally conjugate in $\langle V, V\phi \rangle$.

(iii) now follows from the general result Theorem 4.22. □

Theorem 6.39. Let $\mathfrak{X}$ be a Fitting class of $\mathfrak{S}$-groups. Then the following conditions on the $\mathfrak{S}$-group G are equivalent:

(a) the $\mathfrak{X}$-injectors of G are conjugate,

(b) G has only finitely many $\mathfrak{X}$-injectors,

(c) G has a normal $\mathfrak{X}\mathfrak{S}^*$-subgroup containing all the $\mathfrak{X}$-injectors of G.

Proof. (c) implies (b) and (b) implies (a) are clear.

(a) implies (c). If V is an $\mathfrak{X}$-injector of G, then $Lc\ell(V) = C\ell(V)$ and hence $Lc\ell(VZ) = C\ell(VZ)$. Since G/Z is a $\mathfrak{Y}$-group, Theorem 4.26 shows that $C\ell(VZ)$ is finite. But V is an $\mathfrak{X}$-injector of VZ and $V \triangleleft VZ$. Thus V is the only $\mathfrak{X}$-injector of G contained in VZ and $|C\ell(V)| = |C\ell(VZ)|$ is finite. Therefore $|V^G/V_G|$ is finite and V^G is the normal subgroup described in (c).□

For certain Fitting classes, the injectors can be characterized as the maximal $\mathfrak{X}$-subgroups containing the $\mathfrak{X}$-radical. In particular this is true of the nilpotent injectors of a finite soluble group and this result is easily extended to $\mathfrak{S}$-groups.

Theorem 6.40. The $L\mathfrak{N}$-injectors of the $\mathfrak{S}$-group G are the maximal locally nilpotent subgroups containing the locally nilpotent radical R of G.

Proof. If V is an $L\mathfrak{N}$-injector of G, then V is a maximal locally nilpotent subgroup of G and must contain each normal locally nilpotent subgroup of G; in particular V contains R.

Conversely, let W be a maximal locally nilpotent subgroup of G containing R and let $\{F_i : i \in I\}$ be a local system of finite normal subgroups of G. Then $R \cap F_i$ is the Fitting subgroup of F_i and so $W \cap F_i$ is contained in a nilpotent injector of F_i. Let X_i be the finite non-empty set of nilpotent injectors of F_i which contain $W \cap F_i$. If $F_i \leqslant F_j$ and $V \in X_j$ then $V \cap F_i \in X_i$ so that we can define $\theta_{ji} : X_j \to X_i$ by $V\theta_{ji} = V \cap F_i$. The inverse limit of the system $\{X_i : i \in I\}$ is non-empty and so there is a set $\{V_i : i \in I\}$ such that each V_i is an $\mathfrak{N}$-injector of F_i containing $W \cap F_i$ and, whenever $F_i \leqslant F_j$, we have $V_j \cap F_i = V_i$. If we let $V = \bigcup_{i \in I} V_i$ then $V \cap F_i = V_i$ and, by Lemma 6.35, V is an $L\mathfrak{N}$-injector of G. But $V \geqslant \bigcup_{i \in I} (W \cap F_i) = W$ and, by the maximality of W, we have W = V. □

More detailed results on injectors of $\mathfrak{S}$-groups are given in [36]. In particular, different characterizations are given for $\mathfrak{X}$-injectors for the cases in which $\mathfrak{X}$ is a product of Fitting classes or $\mathfrak{X}$ is locally defined.

7 Centre-by-finite and finite-by-abelian groups

We saw in Chapter 1 that every centre-by-finite group is finite-by-abelian and every finite-by-abelian group is an FC-group. B.H. Neumann has given various characterizations of these classes and it is these results and some generalizations which we present in this Chapter. The proofs given here are slightly different from Neumann's original proofs and we stress the dependence on purely set-theoretic results. The first of these set-theoretic results is perhaps the most basic theorem in combinatorial set theory.

Theorem 7.1. (Ramsey, see [111], p.33) Let S be an infinite set and suppose that the family $[S]^2$ of 2-element subsets of S is expressed as a union of n subfamilies $[S]^2 = \Delta_1 \cup \ldots \cup \Delta_n$, where n is finite. Then there is an infinite subset T of S and a k ($1 \leq k \leq n$) such that $[T]^2 \subseteq \Delta_k$.

Proof. Choose an element $x_1 \in S$; there is a k(1) and an infinite subset S_1 of $S - \{x_1\}$ such that $\{x_1, x\} \in \Delta_{k(1)}$, for all $x \in S_1$. Choose $x_2 \in S_1$; there is a k(2) and an infinite subset S_2 of $S_1 - \{x_2\}$ such that $\{x_2, x\} \in \Delta_{k(2)}$, for all $x \in S_2$. Choosing $x_3 \in S_2$ and continuing the construction we obtain an infinite set of elements $\{x_1, x_2, \ldots\}$ and integers $k(r), 1 \leq k(r) \leq n$, such that $\{x_r, x_s\} \in \Delta_{k(r)}$ if $r < s$. Since the k(r) take only finitely many values, there is an infinite subset $\{x_{r_1}, x_{r_2}, \ldots\}$ such that $k(r_1) = k(r_2) = \ldots = k$, say. This is the required infinite subset T, since each pair $\{x_{r_s}, x_{r_t}\}$ is in Δ_k. □

The generalization of this result to uncountable sets is rather more complicated but we state it here as it can be used to extend some of the results to higher cardinalities.

Theorem 7.2. (Erdös, Hajnal and Rado [22]) Let $\mathfrak{m}$ be an infinite cardinal and S a set with $|S| > \exp \mathfrak{m}$ and suppose that the family $[S]^2$ of 2-element subsets of S is expressed as a union of $\mathfrak{m}$ subfamilies, $[S]^2 = \bigcup_{\alpha<\mu} \Delta_\alpha$.

Then there is an $\alpha < \mu$ *and a subset* T *of* S *such that* $|T| = \mathfrak{m}$ *and* $[T]^2 \subseteq \Delta_\alpha$. □

Our first application of Ramsey's Theorem is to obtain a result about groups which are the union of finitely many cosets.

Lemma 7.3. (Neumann [71]) *Suppose that the group* G *can be expressed as the union of finitely many cosets* $G = \bigcup_{i=1}^{n} H_i g_i$. *If* $H_1 g_1 \not\subseteq \bigcup_{i=2}^{n} H_i g_i$, *then* $|G:H_1|$ *is finite.*

Proof. Let $g \in H_1 g_1 - \bigcup_{i=2}^{n} H_i g_i$. Then $H_1 g_1 = H_1 g$ and

$$G = G\, g^{-1} = H_1 \cup \bigcup_{i=2}^{n} H_i g_i g^{-1}.$$

If $|G:H_1|$ is infinite then we can choose an infinite set $S = \{x_1, x_2, \ldots\}$ of distinct coset representatives of H_1 in G. If $r < s$, then $x_r x_s^{-1} \notin H_1$ and so there is a least $m = m(r,s)$, $2 \leq m \leq n$, such that $x_r x_s^{-1} \in H_m g_m g^{-1}$. We define the subsets Δ_m of $[S]^2$ by $\Delta_m = \{\{x_r, x_s\} : m(r,s) = m\}$. By Ramsey's Theorem there is an infinite subset T of S and a k such that $2 \leq k \leq n$ and $[T]^2 \subseteq \Delta_k$. Let x_r, x_s, x_t be three elements of T with $r < s < t$, so that $g_k g^{-1} x_s x_r^{-1}$, $g_k g^{-1} x_t x_s^{-1}$ and $g_k g^{-1} x_t x_s^{-1}$ are elements in H_k. But then

$$g_k g^{-1} = (g_k g^{-1} x_s x_r^{-1})(g_k g^{-1} x_t x_r^{-1})^{-1} (g_k g^{-1} x_t x_s^{-1}) \in H_k$$

contrary to $g \notin H_k g_k$. Hence $|G:H_1|$ is finite. □

Theorem 7.4. (Baer, see Neumann [72]) *A group* G *is centre-by-finite if and only if* G *is a union of finitely many abelian subgroups.*

Proof. If G/Z is finite, then G/Z is the union of finitely many cyclic groups $\langle aZ \rangle$ and so G is the union of the abelian subgroups $\langle a, Z \rangle$.

Conversely, if G is the union of finitely many abelian subgroups then we can choose a minimal set $A_1, \ldots, A_n$ such that $G = A_1 \cup \ldots \cup A_n$. It is clear that $A_r \not\subseteq \bigcup_{s \neq r} A_s$ and so $|G:A_r|$ is finite for each r. Thus $|G:A_1 \cap \ldots \cap A_n|$ is finite. But $A_1 \cap \ldots \cap A_n$ centralizes each A_r and hence is contained in

Z(G). □

Lemma 7.3 can be extended to groups expressed as a union of an infinite set of cosets but is more difficult to apply. For example, if G is a union of countably many cosets (or subgroups) then one can not usually choose a minimal set of cosets whose union is G. However it is possible to apply the generalization of Lemma 7.3 in some situations and one can prove, for example, that if G is a union of $\mathfrak{m}$ abelian subgroups then $|G/Z(G)| \leq \exp \exp \mathfrak{m}$.

Subgroups of a rather special type will play an important role throughout this chapter. These subgroups were introduced by Neumann [73] and hence we use the term N-group.

An N-*group* of cardinality $\mathfrak{m}$ (infinite) is a group generated by elements $a_i, b_i, i \in I$, where $|I| = \mathfrak{m}$ and the generators satisfy the conditions

$$[a_i, a_j] = [b_i, b_j] = [a_i, b_j] = 1, \text{ if } i \neq j,$$
$$[a_i, b_i] = c_i \neq 1.$$

This does not give a presentation of the group as the generators may satisfy further relations. If we have the further restriction that the elements $c_i, i \in I$, are distinct then we call the group an N_1-*group* of cardinality $\mathfrak{m}$.

We shall show that if G/Z(G) is infinite or if G' is infinite then G contains an N-group or an N_1-group and then obtain the main results by considering these groups in more detail.

<u>Lemma 7.5</u>. <u>Let U be a subgroup of finite index in the FC-group G.</u>

(i) <u>If $|U/Z(U)|$ is finite, then $|G/Z(G)|$ is finite.</u>

(ii) <u>If U' is finite, then G' is finite.</u>

<u>Proof.</u> There is a finitely generated normal subgroup F of G such that FU = G.

(i) Since $|G/C_G(F)|$ is finite and $|G:Z(U)|$ is finite, we have $|G:C_G(F) \cap Z(U)|$ is finite. But $C_G(F) \cap Z(U)$ centralizes FU = G and so G/Z(G) is finite.

(ii) We have $G' \leq U'(F \cap G')$. But $F \cap G'$ is contained in the periodic subgroup of F and so is finite, by Theorem 1.7(ii), and hence G' is finite. □

Theorem 7.6. Let G be an FC-group.

(i) If $|G/Z(G)|$ is infinite, then G contains an infinite N-subgroup.

(ii) If G′ is infinite, then G contains an infinite N_1-subgroup.

Proof. We define the generators a_n, b_n (n=1,2,...) inductively. Suppose that we have defined $a_1, b_1, \ldots, a_{n-1}, b_{n-1}$ and let $C = C_G(a_1, b_1, \ldots, a_{n-1}, b_{n-1})$ so that $|G:C|$ is finite. By Lemma 7.5, $C/Z(C)$ is infinite and, in part (ii), C′ is infinite.

Therefore C contains elements a_n, b_n such that $[a_n, b_n] \neq 1$ and, in part (ii), we can choose a_n, b_n so that $[a_n, b_n] \notin \{1, c_1, \ldots, c_{n-1}\}$. □

We shall use the construction of an N-subgroup first to give a characterization of centre-by-finite groups related to Theorem 7.4.

To each group G we may associate a graph $\Gamma = \Gamma(G)$ in which the vertices are the elements of G and the edges are the pairs $\{x,y\}$ which do not commute. Then the smallest number of abelian subgroups required to cover G is the chromatic number $\chi(\Gamma)$ of the graph Γ. Thus G is centre-by-finite if and only if $\chi(\Gamma)$ is finite. If Γ has an infinite complete subgraph then it must have infinite chromatic number. A complete subgraph of $\Gamma(G)$ corresponds to a set of elements of G, no two of which commute.

We therefore consider groups in which each infinite set S of elements contains a commuting pair. The set of pairs in $[S]^2$ can be split into two subsets by defining

$$\{x,y\} \in \Delta_o \text{ if and only if } [x,y] = 1, \text{ and } \Delta_1 = [S]^2 - \Delta_o.$$

Then, applying Ramsey's Theorem, we see that each infinite subset S of G contains an infinite subset T such that $[x,y] = 1$, for all $x,y \in T$.

Theorem 7.7. (Neumann [76] , Faber, Laver and McKenzie [28]) A group G is centre-by-finite if and only if each infinite subset of G contains a commuting pair.

Proof. If G is centre-by-finite then any infinite set S must contain a pair of elements x,y such that $xy^{-1} \in Z(G)$ and hence $[x,y] = 1$.

Suppose then that each infinite set of elements of G contains a commuting pair. We show first that G is an FC-group. Let $x \in G$ and suppose that $|G:C_G(x)|$ is infinite. Let $S = \{x_1, x_2, \ldots\}$ be an infinite set of elements such that $x_i x_j^{-1} \notin C_G(x)$, whenever $i \neq j$. There is an infinite subset T_o of S such that each pair of elements in T_o commute. Now consider the infinite set $\{xy : y \in T_o\}$. This contains an infinite subset in which each pair of elements commute. Thus there is an infinite subset T of T_o such that, for each pair $\{y,z\} \in [T]^2$, $[y,z] = [xy,xz] = 1$. But then

$$xyz^{-1} = (xy)(z^{-1}x^{-1})x = (z^{-1}x^{-1})(xy)x = z^{-1}yx = yz^{-1}x,$$

contrary to $yz^{-1} \notin C_G(x)$.

If $G/Z(G)$ is infinite then, by Lemma 7.6, G contains an infinite N-subgroup $H = \langle a_n, b_n : n=1,2,\ldots \rangle$. Let $x_n = a_1 \ldots a_n b_n$; then there is a pair x_m, x_n such that $[x_m, x_n] = 1$. But if $m < n$, then

$$\begin{aligned}[x_m, x_n] &= b_m^{-1} a_m^{-1} \ldots a_1^{-1} b_n^{-1} a_n^{-1} \ldots a_1^{-1} a_1 \ldots a_m b_m a_1 \ldots a_n b_n \\ &= b_m^{-1} a_m^{-1} b_n^{-1} a_n^{-1} a_m^{-1} a_m b_m a_m a_n b_n \\ &= b_m^{-1} a_m^{-1} b_m a_m \neq 1\end{aligned}$$

contrary to H being an N-subgroup. Hence $G/Z(G)$ is finite. □

A characterization of finite-by-abelian groups is also easily obtained from the existence of N_1-subgroups.

Theorem 7.8. (Neumann [73]) *A group G is finite-by-abelian if and only if there is an integer n such that $|C\ell(x)| \leq n$, for all $x \in G$.*

Proof. If G' is finite then each conjugate of x is in the coset xG' and so $|C\ell(x)| \leq |G'|$.

Conversely, suppose that G' is infinite; then G contains an infinite N_1-subgroup $H = \langle a_n, b_n : n = 1,2,\ldots \rangle$. For each $m \leq n$, $[a_1 \ldots a_n, b_m] = c_m$ and so $|C\ell(a_1 a_2 \ldots a_n)| \geq n$. □

Groups with a bound on the size of the conjugacy classes are usually referred to as BFC-groups. There has been considerable work done on obtaining a bound for $|G'|$ depending on n and we refer the reader to the paper by

P. M. Neumann and Vaughan-Lee [78] for details of this work which is largely concerned with finite groups.

The remaining results in this Chapter are an extension of the results given by B.H. Neumann in [73]. They may be found in a rather more general form in [29] where the restriction to FC-groups is relaxed. By considering only FC-groups here we are able to use a simpler set-theoretic result. This result is closely related to the Erdös-Rado-Marczewski Theorem on Δ-systems (see [111], Chapter 6.2) but we give a direct proof here.

Theorem 7.9. (Stothers and Tomkinson [96]) Let $X_i, i \in I$, be finite sets with $|I| = \mathfrak{m}$, where $\mathfrak{m}$ is an uncountable cardinal. Then there is a subset J of I such that $|J| = \mathfrak{m}$ and $|\bigcup\{X_j \cap X_k : \{j,k\} \in [J]^2\}| < \mathfrak{m}$.

Proof. Observe first that we can make a number of additional assumptions. First we can adjoin distinct elements x_i to the sets X_i, so that $X_i \cap \{x_i\}$ is finite and $(X_i \cup \{x_i\}) \cap (X_j \cup \{x_j\}) = X_i \cap X_j$, for $i \neq j$. Thus we may assume that the sets X_i are distinct and that $X = \bigcup_{i\in I} X_i$ has cardinality $\mathfrak{m}$.

Secondly, we may assume that the set X is well-ordered with order type μ, the least ordinal of cardinality $\mathfrak{m}$, and that each X_i has the induced ordering. For a subset K of I, we put $U(K) = \bigcup\{X_i \cap X_j : \{i,j\} \in [K]^2\}$ and for each positive integer n, we let

$$X_n(K) = \{x \in X : x \text{ is the mth element of } X_k \text{ for some } k \in K,\ m \lesssim n\}.$$

(I) $\mathfrak{m}$ a regular cardinal. Clearly $X = \bigcup_{n=1}^{\infty} X_n(I)$ and so $\mathfrak{m} = \Sigma_{n=1}^{\infty} |X_n(I)|$. Since $\mathfrak{m}$ is a regular uncountable cardinal, there is a least integer n such that $|X_n(I)| = \mathfrak{m}$. Thus there is a subset K of I such that $|K| = \mathfrak{m}$ and the sets $X_i, i \in K$, have the same initial sequence of length $n-1$ but their n th elements are all different. A straightforward induction argument allows us to define a subset $J = \{j(\alpha) : \alpha < \mu\}$ of K so that the n th element of $X_{j(\alpha)}$ is greater than all elements in $\bigcup_{\beta<\alpha} X_{j(\beta)}$. Then the intersection of any pair of sets X_j, X_k with $\{j,k\} \in [J]^2$ is the common initial segment of length $n-1$. Hence $|U(J)| = n-1$ is finite.

(II) $\mathfrak{m}$ a singular cardinal. If ρ is the cofinality of $\mathfrak{m}$ then we can write $\mathfrak{m} = \Sigma_{\gamma<\rho} \mathfrak{m}_\gamma$, with each $\mathfrak{m}_\gamma$ a regular cardinal. The set I may be partitioned into subsets I_γ of cardinality $\mathfrak{m}_\gamma$, so that $I = \bigcup_{\gamma<\rho} I_\gamma$. By case

(I), each I_γ has a subset J_γ of cardinality $\mathfrak{m}_\gamma$ such that $|U(J_\gamma)| = n_\gamma - 1$ is finite. Hence $|\bigcup_{\gamma<\rho} U(J_\gamma)| \leq |\rho| < \mathfrak{m}$. We construct subsets K_γ of J_γ such that $|K_\gamma| = \mathfrak{m}_\gamma$ and $U(\bigcup_{\gamma<\rho} K_\gamma) \subseteq \bigcup_{\gamma<\rho} U(J_\gamma)$. Suppose that we have constructed sets K_δ, for all $\delta < \gamma$, such that whenever $\delta < \varepsilon < \gamma$, $i \in K_\delta$ and $j \in K_\varepsilon$, we have $X_i \cap X_j \subseteq U(J_\varepsilon)$. Then $|\bigcup\{X_i : i \in \bigcup_{\delta<\gamma} K_\delta\}| < \mathfrak{m}_\gamma$, since $\mathfrak{m}_\gamma$ is regular. From the X_i with $i \in J_\gamma$, choose $\mathfrak{m}_\gamma$ of them whose n_γ-th elements are greater than all elements in $\bigcup\{X_i : i \in \bigcup_{\delta<\gamma} K_\delta\}$. If $i \in K_\delta$ for for some $\delta < \gamma$ and $j \in K_\gamma$ then we have $X_i \cap X_j \subseteq U(J_\gamma)$. Thus we can construct the K_γ's for all $\gamma < \rho$. Clearly $J = \bigcup_{\gamma<\rho} K_\gamma$ is the required set. □

One of the applications of this result concerns sets of commutators in an FC-group. If S and T are subsets of a group then the group [S,T] is generated by the set of commutators $\{[s,t] : s \in S, t \in T\}$. If this set is infinite then $|[S,T]|$ is equal to the cardinality of the set of commutators. If the set of commutators is finite, we sometimes have to distinguish between the cardinality of the set of commutators and the subgroup generated by the commutators.

Corollary 7.10. Let G be an FC-group. If S is a subset of G such that $|[S,G]| = \mathfrak{m}$ is uncountable, then there is a subset T of S such that $|[T,G]| = \mathfrak{m}$ but $|[T,T]| < \mathfrak{m}$.

Proof. For each $s \in S$, the set $X_s = \{[s,g] = g \in G\}$ is finite and so we have $\mathfrak{m}$ finite subsets X_s of $[S,G]$. There is a subset S_1 of S such that $|S_1| = \mathfrak{m}$ and there are distinct elements $x_s \in X_s$, $s \in S_1$. The commutator $[s,t]$ is in the intersection $X_s \cap X_t$. By the theorem there is a subset T of S_1 such that $|T| = \mathfrak{m}$ and hence $|[T,G]| = \mathfrak{m}$ but $|\bigcup\{X_s \cap X_t : \{s,t\} \in [T]^2\}| < \mathfrak{m}$. Hence $[T,T] = \langle [s,t] : \{s,t\} \in [T]^2 \rangle$ has cardinality less than $\mathfrak{m}$. □

We also use Theorem 7.9 in other situations where we can associate some finite set with an element of a group; for example, if G is contained in a direct product of groups then we can associate the set of non-trivial components with each element.

Corollary 7.11. Let A_i, $i \in I$, be an uncountable family of finitely generated subgroups of an abelian group A and let $|I| = \mathfrak{m}$. Then there is a subgroup

$B \leq A$ with $|B| < \mathfrak{m}$ and a subset J of I such that $|J| = \mathfrak{m}$ and $\langle A_iB/B : i \in J \rangle = \mathrm{Dr}_{i \in J}(A_iB/B)$.

Proof. Embed A in a divisible group $\mathrm{Dr}_{\lambda \in \Lambda} D_\lambda$ with each D_λ locally cyclic. For each $i \in I$, let $X_i = \mathrm{supp}(A_i) = \{\lambda \in \Lambda : \pi_\lambda(A_i) \neq 1\}$; since A_i is finitely generated, each X_i is a finite set. By Theorem 7.9, there is a subset J of I such that $|J| = \mathfrak{m}$ and $X = \bigcup\{X_i \cap X_j : \{i,j\} \in [J]^2\}$ has cardinality less than $\mathfrak{m}$. Let $B = A \cap \mathrm{Dr}_{\lambda \in X} D_\lambda$; then $|B| < \mathfrak{m}$. Suppose that $a = \Pi_{i \in J} a_i \in B$; then $a_i = a(\Pi_{j \neq i} a_j^{-1})$ and so $\mathrm{supp}\ (a_i)$ $\subseteq X_i \cap (X \cup \bigcup_{j \neq i} X_j) \subseteq X \cup \bigcup_{j \neq i} (X_i \cap X_j) \subseteq X$. Thus $a_i \in B$ and so $\langle A_iB/B : i \in J \rangle = \mathrm{Dr}_{i \in J}(A_iB/B)$. □

The dual characterizations of centre-by-finite and finite-by-abelian groups given by Neumann [73] depend on the finiteness of $|U^G : U|$ and $|U/U_G|$ for each subgroup U of G. We note first the relationships between these cardinalities and that of $C\ell_G(U)$.

Lemma 7.12. Let G be a group satisfying one of the conditions

(a) $C\ell_G(A)$ is finite, for each abelian subgroup A of G,

(b) $|A^G : A|$ is finite, for each abelian subgroup A of G.

Then G is an FC-group.

Proof. (a) Let $x \in G$; then $|G : N_G(\langle x \rangle)|$ is finite. Since $N_G(\langle x \rangle)/C_G(\langle x \rangle)$ is isomorphic to a group of automorphisms of the cyclic group $\langle x \rangle$ it is finite and hence $|G : C_G(x)|$ is finite.

(b) Let $x \in G$; then $|\langle x \rangle^G : \langle x \rangle| = k$ is finite and so $\langle x^G \rangle$ is a finite extension of a cyclic group. It follows that $\langle x^G \rangle$ is finitely generated and so has only finitely many subgroups of index k. Hence the subgroup $\langle x \rangle$ has only finitely many conjugates and the argument of (i) shows that $|G : C_G(x)|$ is finite. □

We shall sometimes need an additional restriction on the group G. We say that an FC-group G is a $\mathfrak{Z}(\mathfrak{m})$-*group* if each subgroup of G having cardinality $\mathfrak{m}$ is a $\mathfrak{Z}$-group. This condition will usually be used to say that if U is a subgroup of G with $|U| < \mathfrak{m}$, then $|G : C_G(U)| < \mathfrak{m}$. If this were not the case then G would have a normal subgroup F of cardinality $\mathfrak{m}$ such that $U \leq F$ and

and $|FC_G(U):C_G(U)| = \mathfrak{m}$. But then we would have $|F:C_F(U)| = \mathfrak{m}$ contrary to F being a $\mathfrak{Z}$-group.

The condition of being a $\mathfrak{Z}(\mathfrak{m})$-group is, of course, no restriction at all if $\mathfrak{m} = \aleph_0$ and results for that case will be valid in any FC-group.

Lemma 7.13. Let U be a subgroup of the FC-group G and let $\mathfrak{m}$ be an infinite cardinal.

(i) If $|U/U_G| < \mathfrak{m}$, then $|U^G:U| < \mathfrak{m}$.

(ii) If $|C\ell_G(U)| < \mathfrak{m}$, then $|U^G:U| < \mathfrak{m}$.

(iii) Suppose that G is a $\mathfrak{Z}(\mathfrak{m})$-group. Then $|U/U_G| < \mathfrak{m}$ if and only if $|C\ell_G(U)| < \mathfrak{m}$.

Proof. (i) If U/U_G is finite, then U has only finitely many conjugates and so U^G/U_G is finite. If U/U_G is infinite, then it follows from Lemma 1.19 that $|U^G/U_G| = |U/U_G|$.

(ii) If $C\ell_G(U)$ is finite, then U/U_G is finite by Lemma 4.21 and so, by part (i), $|U^G:U|$ is finite. If $C\ell_G(U)$ is infinite, let T be a transversal to $N_G(U)$ in G. If $t \in T$, then there is a finitely generated normal subgroup $F_t = \langle t \rangle^G$ of G such that $t^{-1}Ut \leqslant UF_t$ and so $U^G \leqslant U\langle F_t : t \in T\rangle$. Hence $|U^G:U| \leqslant |T| < \mathfrak{m}$.

(iii) Let $|U/U_G| < \mathfrak{m}$. Then it follows from the $\mathfrak{Z}(\mathfrak{m})$-condition that $|G:C_G(U/U_G)| < \mathfrak{m}$ and hence $|G:N_G(U)| < \mathfrak{m}$.

Conversely, suppose that $|G:N_G(U)| < \mathfrak{m}$; then G has a normal subgroup N such that $NN_G(U) = G$ and N is generated by fewer than $\mathfrak{m}$ elements. Hence $|G:C_G(N)| < \mathfrak{m}$ and so $|U:C_U(N)| < \mathfrak{m}$. But $C_U(N) \triangleleft NN_G(U) = G$ and so $|U/U_G| < \mathfrak{m}$. □

It will be noted that part (iii) of Lemma 7.13 is false in both directions if we omit the $\mathfrak{Z}(\mathfrak{m})$-condition. For, in the extraspecial group given as Example 3.8, $|X/X_G| = \aleph_0$ but $|C\ell_G(X)| = \exp \aleph_0$ and, on the other hand, $|C\ell_G(Y)| = \aleph_0$ but $|Y/Y_G| = \exp \aleph_0$.

The converses of parts (i) and (ii) are false even for $\mathfrak{Z}$-groups. For, let G be a countable extraspecial p-group and let X be an elementary abelian subgroup maximal with respect to $X \cap Z = 1$. Then $|X^G:X| = p$, but $|X/X_G| = |C\ell_G(X)| = \aleph_0$.

As the finiteness of $C\ell_G(U)$ and U/U_G are equivalent in an FC-group it would be nice to add the condition "$|A/A_G|$ finite for all A" to Lemma 7.12. However there are many examples of groups satisfying this condition which are not FC-groups. For example, the extension of a quasicyclic p-group by its automorphism which inverts each element.

The main theorems will now be obtained by showing that groups with a large central factor group or a larged derived subgroup involve a large N-group. To obtain this N-group, we first require an extension of Lemma 7.5.

Lemma 7.14. *Let U be a subgroup of the FC-group G such that* $|G:U| < \mathfrak{m}$.

(i) *If* $|U'| < \mathfrak{m}$, *then* $|G'| < \mathfrak{m}$.

(ii) *If G is a* $\mathfrak{Z}(\mathfrak{m})$-*group and* $|U/Z(U)| < \mathfrak{m}$, *then* $|G/Z(G)| < \mathfrak{m}$.

Proof. Because of Lemma 7.5, we may assume that $\mathfrak{m}$ is uncountable. There is a normal subgroup N of G such that $NU = G$ and $|N| < \mathfrak{m}$.

(i) Clearly $G' \leqslant U'N$ and so $|G'| < \mathfrak{m}$.

(ii) Since G is a $\mathfrak{Z}(\mathfrak{m})$-group, $|G:C_G(N)| < \mathfrak{m}$ and hence $|G:Z(U) \cap C_G(N)| < \mathfrak{m}$. But $Z(U) \cap C_G(N) \leqslant C_G(NU) = Z(G)$ and the result follows. □

Part (ii) of the above Lemma is false for a general FC-group as Example 3.8 again shows. For $|G:Y| = \aleph_0$ and Y is abelian but $|G/Z(G)| = \exp \aleph_0$.

Lemma 7.15. *Let G be an FC-group with* $|G'| \geqslant \mathfrak{m}$.

(i) *If G is a* $\mathfrak{Z}(\mathfrak{m})$-*group, then G contains an* N_1-*subgroup of cardinality* $\mathfrak{m}$.

(ii) *If there is no restriction on G, then G has a normal subgroup F with* $|F| < \mathfrak{m}$ *such that G/F contains an* N_1-*subgroup of cardinality* $\mathfrak{m}$.

Proof. Because of Lemma 7.6(ii), we may assume that $\mathfrak{m}$ is uncountable. Let μ be the least ordinal of cardinality $\mathfrak{m}$.

(i) Suppose that we have defined the elements $a_\beta, b_\beta \in G$ for all $\beta < \alpha$, where α is some ordinal less than μ, to satisfy the conditions

$$[a_\beta,a_\gamma] = [b_\beta,b_\gamma] = [a_\beta,b_\gamma] = 1, \text{ if } \beta \neq \gamma,$$
$$[a_\beta,b_\beta] = c_\beta \notin \langle a_\gamma,b_\gamma : \gamma < \beta\rangle.$$

Let $S_\alpha = \langle a_\beta, b_\beta : \beta < \alpha\rangle$ and let $C_\alpha = C_G(S_\alpha)$. Since $G \in \mathfrak{Z}(\mathfrak{m})$ and $|S_\alpha| < \mathfrak{m}$, we have $|G:C_\alpha| < \mathfrak{m}$ and so, by Lemma 7.14(i), $|C'_\alpha| < \mathfrak{m}$. Therefore C_α contains elements a_α, b_α such that

$$[a_\alpha, b_\alpha] = c_\alpha \in C'_\alpha - S_\alpha.$$

Thus we can define elements a_α, b_α for all $\alpha < \mu$ and so construct the N_1-subgroup $\langle a_\alpha, b_\alpha : \alpha < \mu \rangle$.

(ii) By Corollary 7.10, there is a subset T of G such that $|[T,G]| = \mathfrak{m}$ and $|[T,T]| < \mathfrak{m}$. Factoring out $F_1 = [T,T]^G$ and writing G for G/F_1, we have an abelian subgroup A $(= \langle T \rangle F_1/F_1)$ such that $|[A,G]| = \mathfrak{m}$.

We can choose elements $a_i \in A$, $b_i \in G$, where i belongs to some index set I of cardinality $\mathfrak{m}$, such that $[a_i, b_i] = c_i$ and $c_i \neq c_j$, whenever $i \neq j$. By Corollary 7.10, there is a subset $B = \{b_i : i \in I_1\}$ of $\{b_i : i \in I\}$ such that $|B| = \mathfrak{m}$ and $|[B,B]| < \mathfrak{m}$. Factoring out $F_2 = [B,B]^G$ and replacing G by G/F_2 we now have two abelian subgroups $A_1 = \langle a_i : i \in I_1 \rangle$ and $B_1 = \langle b_i : i \in I_1 \rangle$ with the commutators $c_i = [a_i, b_i]$ being all different.

Now let $X_i = [a_i, B_1] \cup [A_1, b_i]$; since G is an FC-group each X_i is a finite set containing c_i. There is a subset J of I_1 such that $|J| = \mathfrak{m}$ and $|X| < \mathfrak{m}$, where $X = \bigcup \{X_j \cap X_k : \{j,k\} \in [J]^2\}$. If $j \neq k$, then $[a_j, b_k] \in X_j \cap X_k \subseteq X$ and so, factoring out X^G, we obtain elements a_j, b_j, $j \in J$, such that

$$[a_j, a_k] = [b_j, b_k] = [a_j, b_k] = 1, \text{ if } j \neq k,$$

$$[a_j, b_j] = c_j \text{ and } c_j \neq c_k, \text{ whenever } j \neq k.$$

Thus $\langle a_j, b_j : j \in J \rangle$ is the required N_1-subgroup of $G/F_1F_2X^G$. □

Theorem 7.16. *An N_1-group $G = \langle a_i, b_i : i \in I \rangle$ of cardinality $\mathfrak{m}$ which is also an FC-group has an abelian subgroup X such that $|X^G : X| = \mathfrak{m}$.*

Proof. (I) $\mathfrak{m}$ uncountable. For each $i \in I$, let $G_i = \langle a_i, b_i \rangle$; then G_i is a finitely generated FC-group and so $Z_i = Z(G_i)$ is a finitely generated subgroup of $Z = Z(G)$. By Corollary 7.11, there is a subgroup Y of Z with $|Y| < \mathfrak{m}$ and a subset J of I with $|J| = \mathfrak{m}$ such that $\langle Z_iY/Y : i \in J \rangle = \mathrm{Dr}_{i \in J}(Z_iY/Y)$. Now $c_i \in Z_i$ and the c_i's are all different. Since $|Y| < \mathfrak{m}$, we can therefore find a subset K of J such that $|K| = \mathfrak{m}$ and Z_iY/Y is nontrivial for each $i \in K$.

Let $x \in G_iY \cap \langle G_jY : j \neq i, j \in K \rangle$; then we can write $x = g_iy = gy_1$, with $g_i \in G_i$, $g \in \langle G_j : j \neq i \rangle$ and $y, y_1 \in Y$. Thus $g_i = gy_1y^{-1} \in Z(G_i) = Z_i$ and $g = g_iyy_1^{-1} \in Z(\langle G_j : j \neq i \rangle) = \langle Z_j : j \neq i \rangle$. Hence $x \in Z_iY \cap \langle Z_jY : j \neq i, j \in K \rangle = Y$ and so $\langle G_iY/Y : i \in K \rangle = \mathrm{Dr}_{i \in K}\,(G_iY/Y)$.

Now well-order the index set K and label the elements by ordinals $\alpha, \alpha < \mu$. For each $\alpha < \mu$, the group $G_\alpha G_{\alpha+1}Y/Y$ is a direct product of two nonabelian groups and so $G_\alpha G_{\alpha+1}Y$ contains a cyclic group X_α such that $X_\alpha Y/Y$ is not normal in $G_\alpha/G_{\alpha+1}\,Y/Y$. Let

$$X = \langle Y, X_\alpha : \alpha = \lambda + 2n, \text{ where } \lambda \text{ is a limit ordinal and } n \text{ a natural number} \rangle.$$

Then $|X^G : X| = \mathfrak{m}$, as required.

(II) $\mathfrak{m}$ <u>countable</u>. In this case we do not have Corollary 7.11 available and so the proof is considerably more complicated.

We write $G = \langle a_n, b_n : n = 1,2,\ldots \rangle$ where the commutators $c_n = [a_n, b_n]$ are all different. Let $A = \langle a_n : n = 1,2,\ldots \rangle$ and $B = \langle b_n : n = 1,2,\ldots \rangle$. If there were infinitely many integers n such that $[a_n, c_n] \neq 1$, then for two such integers m,n we would have $[a_n, c_n c_m^{-1}] = [a_n, c_n] \neq 1$ so that $c_n c_m^{-1} \notin A$. Since $A^G \geq \langle a_n, c_n : n=1,2,\ldots \rangle$ it would follow that $|A^G : A|$ is infinite. Similarly if there were infinitely many integers n such that $[b_n, c_n] \neq 1$, then we would have $|B^G : B|$ infinite. By removing the finite set of integers n for which $[a_n, c_n] \neq 1$ or $[b_n, c_n] \neq 1$ we may therefore assume that $[a_n, c_n] = [b_n, c_n] = 1$, for all n, and so G is nilpotent of class two.

There is a maximal torsion-free subgroup F contained in the centre of G. It is sufficient to show that $\bar{G} = G/F$ contains an abelian subgroup $\bar{X} = X/F$ such that $|\bar{X}^{\bar{G}} : \bar{X}|$ is infinite. For, $X' \leq F$ and since X' is periodic, X must be abelian and $|X^G : X| = |\bar{X}^{\bar{G}} : \bar{X}|$ is infinite. We may therefore assume that G is periodic.

The subgroup $C = \langle c_n : n=1,2,\ldots \rangle$ is therefore a countable periodic abelian group. Suppose that C contains a subgroup Q of type C_{p^∞}. If $Q \leq A$, then $A = Q \times A_1$ and $A_1^G \geq Q$ so that $|A_1^G : A_1|$ is infinite. If $Q \not\leq A$, then $A^G \geq Q$ and $A \cap Q$ is finite so that $|A^G : A| = |AQ : A| = |Q : A \cap Q|$ is infinite. Therefore C is reduced and so is a direct product of cyclic groups.

We can now choose elements $d_n \in A$, $e_n \in B$ such that

$$[d_m, e_n] = 1, \text{ if } m \neq n,$$
$$[d_n, e_n] = f_n \text{ has prime order } p_n,$$
$$f_m \neq f_n \text{ if } m \neq n$$
$$d_n, e_n \text{ are } p_n\text{-elements.}$$

Suppose we have defined $d_1, e_1, \ldots, d_{n-1}, e_{n-1}$. Then there is an integer k such that $\langle d_1, e_1, \ldots, d_{n-1}, e_{n-1} \rangle \leqslant \langle a_1, b_1, \ldots, a_k, b_k \rangle$. Let $C_1 = \langle c_{k+1}, \ldots \rangle$; then Soc (C_1) is infinite and so there is an element $c \in C_1 - \langle a_1, b_1, \ldots, a_k, b_k \rangle$ of prime order p_n. We let $f_n = c$. Since G is nilpotent, we can choose p_n-elements $d_n \in \langle a_{k+1}, \ldots \rangle$ and $e_n \in \langle b_{k+1}, \ldots \rangle$ such that $[d_n, e_n] = f_n$.

We now consider the two cases in which (a) there are infinitely many distinct primes p_n or (b) the prime p occurs infinitely often as a p_n.

In the first case, by choosing a suitable subset, we may assume that the p_n are all different and are odd. Let $H_n = \langle d_n, e_n \rangle$; then H_n contains elements x_n, y_n such that $[x_n, y_n] = f_n$ but $f_n \notin \langle x_n \rangle$. For, otherwise f_n would be contained in each subgroup generated by a non-central element. If $Z(H_n) = C \times D$, where C is a cyclic group containing f_n then in H_n/D, $\langle \bar{f}_n \rangle$ would be the unique subgroup of order p_n. This can not occur in a non-abelian group if p_n is odd. Now let $X = \langle x_1, x_2, \ldots \rangle$; then X is abelian, $f_n \in X^G$ and $f_m f_n^{-1} \notin X$, if $m \neq n$. For if $f_m f_n^{-1} \in X$, then taking the p_n-th power we obtain $f_m \in X$ and the Sylow p_m-subgroup of X is $\langle x_m \rangle$ which does not contain f_m. Since $f_m f_n^{-1} \notin X$, it follows that $|X^G : X|$ is infinite.

In the second case, by again choosing a suitable subset of the generators, we may assume that $G = \langle d_n, e_n : n=1,2,\ldots \rangle$ is a p-group and $[d_n, e_n] = f_n$ has order p. Thus $D^p \leqslant Z(G)$ and the elements d_n belong to different cosets of Z(G). Hence $D/D^p = \mathrm{Dr}_{n=1}^{\infty} \langle \bar{d}_n \rangle$, where $\bar{d}_n = d_n D^p$.

We now choose elements $x_n \in D$, $y_n \in E$ such that

$$[x_m, y_n] = 1, \text{ if } m \neq n$$
$$[x_n, y_n] = z_n,$$
$$z_m \neq z_n, \text{ if } m \neq n,$$
$$\langle x_1, x_2, \ldots \rangle = \mathrm{Dr}_{n=1}^{\infty} \langle x_n \rangle .$$

Suppose that we have defined the elements $x_1, y_1, \ldots, x_{n-1}, y_{n-1}$. The finite subgroup $<x_1, \ldots, x_{n-1}>$ is a pure subgroup of D and so is contained in a finite direct factor X_1 of D, say $D = X_1 \times X_2$. Now there is an integer k such that, in the group D/D^p we have $<\bar{x}_1, \ldots, \bar{x}_{n-1}> \leq Dr_{i=1}^{k} <\bar{d}_i>$. Since X_2 is infinite, so is $X_2 \cap C_G(y_1, \ldots, y_{n-1})$. Therefore there is an element $x \in X_2 \cap C_G(y_1, \ldots, y_{n-1})$ such that $\bar{x}$ has a non-trivial component in $<\bar{d}_i>$, for some $i > k$. We can define $x_n = x$ and $y_n = e_i$ to complete our inductive definition.

Since Z is an abelian group of exponent p, we can write $Z = (X \cap Z) \times T$ and hence $XZ = X \times T$. If T is infinite, then $|X^G:X| = |XZ:X|$ is infinite and so we may assume that T is finite. In this case $T \leq <x_1, y_1, \ldots, x_k, y_k>$, for some k, and by factoring out this subgroup we may assume that $Z \leq X$.

Finally we define elements u_n, v_n, such

$$[u_m, u_n] = [v_m, v_n] = [u_m, v_n] = 1, \text{ if } m \neq n,$$

$$[u_n, v_n] = w_n \neq 1$$

$$w_m w_n^{-1} \notin <u_n : n=1,2,\ldots>, \quad \text{if } m \neq n .$$

Since $U^G \geq <w_n : n=1,2,\ldots>$ it will then follow that $|U^G:U|$ is infinite. Suppose that we have defined $u_1, v_1, \ldots, u_{n-1}, v_{n-1}$ satisfying the above conditions with $w_k w_\ell^{-1} \notin <u_1, \ldots, u_{n-1}>$ for any $k < \ell < n$. There is an integer t such that $X \cap <u_1, v_1, \ldots, u_{n-1}, v_{n-1}> \leq Dr_{i=1}^{t} <x_i>$. There are integers $r(>t)$ such that z_r has a non-trivial component in some $<x_j>$ with $j > t$. Choose r such that $|<x_r>|$ is minimal. There is now an integer t' such that $X \cap <u_1, v_1, \ldots, u_{n-1}, v_{n-1}, x_r, z_r> \leq Dr_{i=1}^{t'} <x_i>$. There is an integer $s(>t)$ such that z_s has a non-trivial component in $<x_j>$ for some $j > t'$. By the choice of r, $|<x_r>| \leq |<x_s>|$ and so any non-trivial power of $x_r x_s$ will have a non-trivial component in $<x_s>$.

Let $u_n = x_r x_s$ and $v_n = y_r$ so that $w_n = [u_n, v_n] = z_r$. If $m < n$, then $[u_m, v_n] = [u_n, v_m] = 1$. Since w_n has a component in $<x_j>$ for some $j > t$ but does not have a component in $<x_s>$, we have

$$w_m w_n^{-1} \notin Dr_{i=1}^{t} <x_i> \times <x_r x_s>$$

and hence $w_m w_n^{-1} \notin <u_1, \ldots, u_n>$.

Also if $\ell < m < n$, then $w_\ell w_m^{-1} \in Dr_{i=1}^{t} <x_i>$ and so if $w_\ell w_m^{-1} \in <u_1, \ldots, u_n> =$

$= \langle u_1,\ldots,u_{n-1}\rangle \times \langle x_r x_s\rangle$, it would follow that $w_\ell w_m^{-1} \in \langle u_1,\ldots,u_{n-1}\rangle$ contrary to hypothesis. Thus the elements u_n, v_n satisfy all the conditions necessary for the inductive definition.

This completes the proof of Theorem 7.16. □

Combining this Theorem with Lemmas 7.15 and 7.13 we obtain the following characterization.

Theorem 7.17. (Neumann [73], Eremin [23], Tomkinson [107]).

(i) A group G is finite-by-abelian if and only if $|A^G:A|$ is finite for each abelian subgroup A of G.

(ii) Let G be an FC-group and $\mathfrak{m}$ an infinite cardinal. Then $|G'| < \mathfrak{m}$ if and only if $|U^G:U| < \mathfrak{m}$ for each subgroup U of G.

(iii) Let G be a $\mathfrak{Z}(\mathfrak{m})$-group. Then $|G'| < \mathfrak{m}$ if and only if $|A^G:A| < \mathfrak{m}$ for each abelian subgroup A of G. □

The following example shows that it is not sufficient to consider abelian subgroups in part (ii) of the above Theorem.

Example 7.18. Let $X = \langle x_i : i \in I\rangle$ and $Y = \langle y_i : i \in I\rangle$ be two Ehrenfeucht-Faber p-groups of cardinality $\mathfrak{m}$ with centres X_o and Y_o, respectively. Let $Z = \mathrm{Dr}_{i\in I}\langle z_i\rangle$, where z_i has order p and let G be the split extension of $(X \times Z)$ by Y in which

$$[x_i, y_j] = \begin{cases} 1, & \text{if } i \neq j,\\ z_i, & \text{if } i = j.\end{cases}$$

Then $Z(G) = X_o \times Z \times Y_o = G'$.

Suppose U is a subgroup such that $|UG'/G'| = \mathfrak{m}$. Then the projection of U on either XG'/G' or YG'/G' has cardinality $\mathfrak{m}$ and so U is nonabelian. Thus every abelian subgroup A satisfies the condition $|AG'/G'| < \mathfrak{m}$ and hence $|A/(A \cap G')| < \mathfrak{m}$. But $A \cap G' \lhd G$ and so $|A/A_G| < \mathfrak{m}$. By Lemma 7.13(i) it follows that $|A^G:A| < \mathfrak{m}$ for each abelian subgroup A although $|G'| = \mathfrak{m}$. □

The dual result about the index of Z(G) can now be obtained fairly easily using the results about G' and the following elementary result about abelian groups.

Lemma 7.19. *Let* B *be a subgroup of the abelian group* A *such that* $|B| < \mathfrak{m}$. *Then there is a subgroup* C *of* A *such that* $B \cap C = 1$ *and*

(i) *if* $\mathfrak{m}$ *is uncountable,* $|A/C| < \mathfrak{m}$.

(ii) *if* $\mathfrak{m} = \aleph_0$, A/C *satisfies the minimal condition.*

Proof. (i) Embed A in its divisible hull $D = Dr_{i \in I} D_i$, where each D_i is locally cyclic and hence countable. Since $|B| < \mathfrak{m}$ the set $J = supp(B)$ has cardinality less than $\mathfrak{m}$. Let $C = A \cap Dr_{i \in I-J} D_i$; then $B \cap C = 1$ and A/C, being isomorphic to a subgroup of $Dr_{i \in J} D_i$, has cardinality less than $\mathfrak{m}$.

(ii) For each $b \in B$, choose a subgroup M_b maximal subject to $b \notin M_b$. Then A/M_b is either a finite cyclic group or of type C_{p^∞}. Since B is finite $A/\bigcap_{b \in B} M_b$ satisfies the minimal condition and $B \cap \bigcap_{b \in B} M_b = 1$. □

Theorem 7.20. (Neumann [73], Eremin [23], Tomkinson [107]) *Let* G *be a* $\mathfrak{Z}(\mathfrak{m})$*-group. Then the following conditions are equivalent:*

(a) $|G/Z(G)| < \mathfrak{m}$,

(b) $|A/A_G| < \mathfrak{m}$ *for all abelian subgroups* A *of* G,

(c) $|C\ell_G(A)| < \mathfrak{m}$ *for all abelian subgroups* A *of* G.

Proof. The equivalence of (b) and (c) was proved in Lemma 7.13(iii). If $|G/Z| < \mathfrak{m}$, then it is clear that $|A/A_G| \leq |A/(A \cap Z)| < \mathfrak{m}$, for all subgroups A of G. We therefore assume that $|G/Z| \geq \mathfrak{m}$ and show that G must have an abelian subgroup X such that $|X/X_G| \geq \mathfrak{m}$.

If $|G'| \geq \mathfrak{m}$, then it follows from Theorem 7.17 that G has an abelian subgroup X such that $|X^G : X| \geq \mathfrak{m}$ and hence, by Lemma 7.13(i), $|X/X_G| \geq \mathfrak{m}$. We may therefore assume that $|G/Z| \geq \mathfrak{m}$ and $|G'| < \mathfrak{m}$.

We now show that G contains an N-group $\langle a_\alpha, b_\alpha : \alpha < \mu \rangle$ of cardinality $\mathfrak{m}$. Suppose that we have defined the elements a_β, b_β for all $\beta < \alpha$ such that

$$[a_\beta, a_\gamma] = [b_\beta, b_\gamma] = [a_\beta, b_\gamma] = 1, \text{ if } \beta \neq \gamma,$$

$$[a_\beta, b_\beta] = c_\beta \neq 1.$$

Let $S_\alpha = \{a_\beta, b_\beta : \beta < \alpha\}$ and let $C_\alpha = C_G(S_\alpha)$. Since G is a $\mathfrak{Z}(\mathfrak{m})$-group, we have $|G:C_\alpha| < \mathfrak{m}$ and so, by Lemma 7.14(ii), $|C_\alpha/Z(C_\alpha)| \geq \mathfrak{m}$. Therefore there are elements $a_\alpha, b_\alpha \in C_\alpha$ such that $[a_\alpha, b_\alpha] = c_\alpha \neq 1$. Thus we can define the a_α, b_α for all $\alpha < \mu$.

The subgroup $A = \langle a_\alpha : \alpha < \mu \rangle$ is abelian, $|A/(A \cap Z)| = \mathfrak{m}$ and $|A \cap G'| < \mathfrak{m}$.

If $\mathfrak{m}$ is uncountable, then by Lemma 7.19, there is a subgroup X of A such that $|A/X| < \mathfrak{m}$ and $X \cap G' = X \cap (A \cap G') = 1$. It follows that $|X/(X \cap Z)| = \mathfrak{m}$ and also $[X_G, G] \leq X \cap G' = 1$ so that $X_G = X \cap Z$. Hence $|X/X_G| = \mathfrak{m}$.

If $\mathfrak{m} = \aleph_0$, then there is a subgroup X of A such that $X \cap G' = 1$ and A/X satisfies the minimal condition. If $X/(X \cap Z)$ were finite, then $A/(X \cap Z)$ and hence $A/(A \cap Z)$ would satisfy the minimal condition. But $A/(A \cap Z) \cong AZ/Z$ is residually finite and so it would follow that $A/(A \cap Z)$ is finite. This contradiction shows that $X/(X \cap Z)$ is infinite and, as in the uncountable case, $X_G = X \cap Z$ so that $|X/X_G|$ is infinite. □

In the Ehrenfeucht-Faber p-group G of cardinality $\mathfrak{m}$ any subgroup of cardinality $\mathfrak{m}$ contains G′ and so is normal. Thus $|U/U_G| < \mathfrak{m}$ for every subgroup U of G even though $|G/Z(G)| = \mathfrak{m}$.

However, all the abelian subgroups A of G which are maximal subject to $A \cap G' = 1$ have $\mathfrak{m}$ conjugates and so this group does not provide an answer to the following question.

<u>Question 7A.</u> Is there an FC-group G with $|G/Z(G)| = \mathfrak{m}$ but $|C\ell_G(A)| < \mathfrak{m}$ for all (abelian) subgroups A of G?

The condition $|C\ell_G(A)| < \mathfrak{m}$ can of course be written as $|G:N_G(A)| < \mathfrak{m}$ and the hypothesis of Theorem 7.20 holds if we have the condition $|G:C_G(A)| < \mathfrak{m}$ for all subgroups A. Thus Question 7A is related to the questions in Chapter 3 concerned with the classes $\mathfrak{Y}$ and $\mathfrak{Z}$.

8 Miscellaneous topics

FL-Groups

A group G is said to be an FL-*group* (or to have finite layers) if there are only finitely many elements of order n, for each positive integer n or for n infinite. It is immediately clear that an FL-group is periodic and is an FC-group. The class of FL-groups is completely characterized in the following result.

Theorem 8.1. (Černikov [17], Polovickiĭ [81, 82]) The following conditions on a group G are equivalent:

(a) G is an FL-group,

(b) G is a periodic FC-group and, for each prime p, the Sylow p-subgroups of G are centre-by-finite Černikov groups,

(c) G is a periodic FC-group and, for each prime p, $G/O_{p'}(G)$ is a centre-by-finite Černikov group,

(d) G is isomorphic to a subgroup of a direct product $\mathrm{Dr}_{n=1}^{\infty} X_n$, where each X_n is a centre-by-finite Černikov group and, for each prime p, only finitely many of the X_n have elements of order p.

Proof. (a) implies (b). Let P be a Sylow p-subgroup of G and let A be a maximal abelian normal subgroup of P, so that $C_P(A) = A$. Then A is an abelian p-group of finite rank and so is the direct product of a divisible group D and a finite group B. By Theorem 1.9, $D \leq Z(P)$ and so $A = C_P(A) = C_P(B)$ has finite index in P. Hence P/D is finite and P is a centre-by-finite Černikov group.

(b) implies (c). Let P be a Sylow p-subgroup of G and $C = C_G(P)$. Since P is a Černikov group, P = DF for some divisible group D and finite group F. Since $D \leq Z(G)$, $C = C_G(P) = C_G(F)$ has finite index in G. It is clear that $C \cap P$ is a normal Sylow p-subgroup of C and so, by the Schur-Zassenhaus Theorem 5.25, C has a Sylow p'-subgroup Q such that $C = Q(C \cap P)$; clearly Q is normal in C. Now G/C_G is finite and $Q \cap C_G$ is a normal Sylow

p'-subgroup of C_G and hence $Q \cap C_G \leq O_{p'}(G)$. Also $C_G/(Q \cap C_G)$ is isomorphic to a subgroup of P and so is a centre-by-finite Černikov group. Hence $G/(Q \cap C_G)$ is also a centre-by-finite Černikov group.

(c) implies (d). Let $M_p = O_{p'}(G)$, for each prime p. Then clearly $\bigcap_p M_p = 1$ and there is a natural embedding of G into the cartesian product $\Pi_p (G/M_p)$. If F is a finite normal subgroup of G, then for all but finitely many primes p, F is a p'-group and hence $F \leq M_p$. Thus G embeds into the direct product $Dr_p (G/M_p)$. Also each element of order p is contained in P^G, where P is a Sylow p-subgroup of G. Since G is p'-by-Černikov, P is a Černikov group and so $P = DF$ where F is finite and D is divisible. Since $D \leq Z(G)$, we have $P^G = DF^G$ and since F^G is finite, P^G is contained in M_q for all but finitely many primes q, and so there are only finitely many G/M_q which contain elements of order p.

(d) implies (a). The elements of order n are contained in the product of finitely many X_i and this product is itself a centre-by-finite Černikov group H, say. Now H has a finite normal subgroup N such that H/N is divisible abelian of finite rank. It is clear that the elements of order dividing n in H/N form a finite subgroup K/N and so the elements of order n in H are contained in the finite subgroup K. □

It is somewhat surprising to observe that the *Q*-closure of the class of FL-groups is not immediate from the definition but is a trivial consequence of either of the equivalent conditions (b) and (c) above.

Corollary 8.2. *If N is a normal subgroup of the FL-group, then G/N is also an FL-group.* □

FL-groups can be thought of as those groups in which for each integer n, or for n infinite, there are only finitely many cyclic subgroups of order n. One can restrict the class further by considering groups in which each isomorphism class of (abelian) subgroups is finite. These groups have the following characterization.

Theorem 8.3. (Fuchs [34]) *The following conditions on a group G are equivalent:*

(a) each set of isomorphic subgroups of G is finite,

(b) each set of isomorphic abelian subgroups of G is finite,

(c) G is a finite extension of a central subgroup K which is periodic and locally cyclic.

Proof. (a) implies (b) is obvious.

(b) implies (c). It follows from Theorem 7.20 that G/Z(G) is finite. Clearly Z has finite p-rank for each prime p and there are only finitely many primes for which the rank is greater than one, since the group $\mathrm{Dr}_p\,(C_p \times C_p)$ has infinitely many subgroups isomorphic to $\mathrm{Dr}_p\, C_p$. Also the group $C_{p^\infty} \times C_{p^\infty}$ contains infinitely many subgroups isomorphic to C_{p^∞} and so the p-subgroups of Z with rank greater than one are finite extensions of C_{p^∞} or are finite. Hence Z is a finite extension of a periodic locally cyclic group.

(c) implies (a). There is a finite normal subgroup F of G such that FK = G. Let $|F| = m$ and let K_1 be the cyclic subgroup of K consisting of all elements of order dividing m. Then $N = FK_1$ is the set of all elements of G having order dividing m. For, if g = fk then $g^m = f^m k^m = k^m$ and so $g^m = 1$ if and only if $k \in K_1$. Now let X be a fixed subgroup of G and Y a subgroup of G isomorphic to X. Then $X \cap N$ is the set of elements of X having order dividing m and hence $X/(X \cap N) \cong Y/(Y \cap N)$. Hence XN/N and YN/N are isomorphic subgroups of the locally cyclic group G/N and so XN = YN. Thus $Y \leqslant XN$ and $|XN:Y| \leqslant |N| = n$, say. Thus $|XN \cap K : Y \cap K| \leqslant n$ and so $Y \cap K \geqslant (XN \cap K)^n$. Therefore every subgroup Y isomorphic to X satisfies $(XN \cap K)^n \leqslant Y \leqslant XN$. Since $XN/(XN \cap K)^n$ is finite, there are only finitely many subgroups of G isomorphic to X.

Infinite Abelian and Nilpotent Subgroups

It is well known that every infinite locally finite group has an infinite abelian subgroup ([84], Part 1, p.95) and this can be proved very easily for FC-groups (see Lemma 8.4 below). Faber and Semenova have considered the existence of abelian or nilpotent subgroups having the same cardinality as an FC-group G. It was this question which led to the construction of the Ehrenfeucht-Faber group in Theorem 3.12. We give below the positive results on the existence of large subgroups and also see to what extent the Ehrenfeucht-Faber group is typical of FC-groups without abelian subgroups of the same cardinality as the whole group.

Lemma 8.4. A maximal abelian subgroup A of the infinite FC-group G is infinite and $\exp|A| \geq |G|$.

If G is a $\mathfrak{Z}$-group, then $|A| = |G|$.

Proof. We have $C_G(A) = A$ and so, if A is finite, $|G:A|$ is finite and we obtain $|G| = |G:A|\,|A|$ is finite. Since A is infinite, Lemma 1.19 shows that $|A^G| = |A|$ and so $G/C_G(A^G)$, being isomorphic to a group of automorphisms of A^G, has cardinality at most $\exp|A|$. Therefore $|G| \leq |G:C_G(A^G)|\,|A| \leq \exp|A|$.

If G is a $\mathfrak{Z}$-group, then $|G:A| = |G:C_G(A)| \leq |A|$ so that $|G| = |A|$. □

Theorem 8.5. (Semenova [89]) Let N be a maximal class two nilpotent subgroup of the infinite FC-group G. Then $|N| = |G|$.

Proof. Suppose $|N| < |G|$; then since $Z(G) \leq N$, we have $|G/Z| = |G|$. Since G/Z is a $\mathfrak{Z}$-group, $|G:C_G(N/Z)| \leq |N/Z| < |G|$ and so $|C_G(N/Z)/Z| = |G|$. Let A/Z be a maximal abelian subgroup of $C_G(N/Z)/Z$; then by Lemma 8.4, $|A/Z| = |G|$ and hence $|AN| = |G|$. We obtain a contradiction by showing that AN is nilpotent of class two and hence N = AN.

Now $[N,A] \leq Z$ and so $[N,A,N] = [A,N,N] = 1$. Therefore, by the Three Subgroups Lemma ([84], Part 1, p.44), we have $[N',A] = 1$. But $(AN)' = [A,N]\,N' \leq ZN'$ and so $[(AN)',AN] \leq [ZN',AN] = [N',A]\,[N',N] = 1$. Therefore AN is nilpotent of class two. □

Lemma 8.5. Let G be an infinite residually finite periodic FC-group. Then $|Soc(G)| = |G|$.

Proof. Let S = Soc(G) and, for each $x \in S$, choose a normal subgroup N_x of finite index in G such that $x \notin N_x$. Then $\bigcap_{x\in S} N_x = 1$ and so, by Lemma 2.22, $|G/Z| \leq |S|$. But $Soc(Z) = S \cap Z$ and since Z is a residually finite periodic abelian group, $|Soc(Z)| = |Z|$. Therefore $|Z| \leq |S|$ and so $|G| = |G/Z|\,|Z| \leq |S|$. □

Lemma 8.6. Let A be a maximal abelian normal subgroup of the infinite locally soluble residually finite periodic FC-group G. Then $|A| = |G|$.

Proof. Let M be a minimal normal subgroup of G; then M is abelian. If $M \not\leq A$, then $M \cap A = 1$ so that $[M,A] = 1$ and MA is abelian. Thus A contains all minimal normal subgroups of G and so $A \geq Soc(G)$. The result now follows from Lemma 8.5. □

Theorem 8.7. (Semenova [89]) Let N be a maximal class two nilpotent normal subgroup of the infinite locally soluble FC-group G. Then $|N| = |G|$.

Proof. Suppose $|N| < |G|$; then since $Z(G) \leq N$, we have $|G/Z| = |G|$. Since G/Z is a $\mathfrak{Z}$-group, we obtain as in Theorem 8.3 that $|C_G(N/Z)/Z| = |G|$. Let $S/Z = Soc(C_G(N/Z)/Z)$; then since $C_G(N/Z)/Z$ is periodic and residually finite, Lemma 8.5 shows that $|S/Z| = |G|$. Also the socle is a characteristic subgroup and so $S \lhd G$. The argument of Theorem 8.3 shows that SN is a normal class two nilpotent subgroup of G and hence N = SN. □

Theorem 8.8. (Ehrenfeucht and Faber [20]) An infinite (locally soluble) FC-group G has a (normal) subgroup H such that $|H| = |G|$ and either (a) H is torsion-free abelian, or (b) H is periodic and nilpotent of class two.

Proof. Let N be a maximal (normal) class two nilpotent subgroup of G so that by Theorems 8.3 and 8.7, $|N| = |G|$. Let T be the subgroup of periodic elements of N; then T is characteristic in N and if $|T| = |N|$, we have case (b).

So suppose that $|T| < |N|$ and let $Z = Z(G) \leq N$. There is a torsion-free subgroup H of Z such that Z/H is periodic and hence N/H is periodic. Clearly $H \cap T = 1$ and so $HT/T \cong H$. Thus HT/T is a subgroup of the uncountable torsion-free abelian group N/T such that N/HT is periodic and so $|HT/T| = |N/T|$. (This can be seen, for example, by considering the embedding of N/T in its divisible hull $Dr_{i \in I} D_i$. HT/T must have a non-trivial projection on each D_i and so $|HT/T| = |I| = |N/T|$). It follows that $|H| = |HT/T| = |N/T| = |G|$, as required. □

It follows from this last result that an FC-group G without abelian subgroups of the same cardinality as G must contain a periodic class two nilpotent subgroup N such that $|N| = |G|$ and, of course, N has no abelian subgroups of the same cardinality as N.

We can go a little further than this. Since it is nilpotent, N is the direct product of its Sylow p-subgroups N_p and $|N| = \sum_p |N_p|$ is uncountable If each N_p has an abelian subgroup A_p of cardinality $|N_p|$ then $A = \mathrm{Dr}\, A_p$ is an abelian subgroup of N and $|A| = \sum_p |A_p| = |N|$. Therefore G has a p-subgroup P which is nilpotent of class two and has no abelian subgroups of cardinality $|P|$. Since $Z(P)$ is abelian, $|P/Z(P)| = |P|$ and also $P/Z(P)$ is residually finite and abelian; therefore $|\mathrm{Soc}(P/Z(P))| = |P/Z(P)| = |P|$. Let $S/Z(P) = \mathrm{Soc}(P/Z(P))$; then $|S| = |P|$, S has no abelian subgroup of cardinality $|S|$, $S/Z(S)$ is elementary abelian and hence S' is elementary abelian.

Let $F_i, i \in I$, be the finite subgroups of S' so that $|I| < |S|$. For each element $x \in S$, the commutator $[x,S]$ is one of the finite subgroups F_i and so $S = \bigcup_{i \in I} C_S(S/F_i)$. If $|S|$ is regular, then there is an $i \in I$ such that $|C_S(S/F_i)| = |S|$. Let $X = C_S(S/F_i)$; then $|X| = |S|$, X' is contained in F_i and so is finite and elementary abelian and X has no abelian subgroup of cardinality $|X|$.

If $|S|$ is singular, then we need to assume G C H (or, more precisely, if $\mathfrak{m} < |S|$ then $\exp \mathfrak{m} < |S|$). We can choose a set of subgroups $X_\alpha = C_S(S/F_\alpha)$, $\alpha < \gamma$ = cofinality of $\mathfrak{m}$, such that $|X_\alpha| = \mathfrak{m}_\alpha$ and $\sum_{\alpha<\gamma} \mathfrak{m}_\alpha = |S|$ with $\mathfrak{m}_\alpha < \mathfrak{m}_\beta$, whenever $\alpha < \beta$. Suppose that X_α has an abelian subgroup A_α of cardinality $\mathfrak{m}_\alpha$. Let $C_\alpha = C_S(A_\alpha)$; then $|S:C_\alpha| \leq \exp \mathfrak{m}_\alpha < |S|$ and so there is a $\beta > \alpha$ such that $|S:C_\alpha| < \mathfrak{m}_\beta$. Thus $|X_\beta \cap C_\alpha| = \mathfrak{m}_\beta$. If $X_\beta \cap C_\alpha$ has no abelian subgroup of cardinality $\mathfrak{m}_\beta$ we take $X = X_\beta \cap C_\alpha$. If there is an abelian subgroup B_β of $X_\beta \cap C_\alpha$ of cardinality $\mathfrak{m}_\beta$ then we let $A_\beta = A_\alpha B_\beta$. Then A_β is an abelian subgroup of cardinality $\mathfrak{m}_\beta$ and we consider $X_\delta \cap C_S(A_\beta)$ for an appropriate $\delta > \beta$. In this way we obtain an ascending chain of abelian subgroups A_β of cardinality $\mathfrak{m}_\beta$. The union of this chain would be an abelian subgroup of cardinality $\mathfrak{m}$. Thus there is a subgroup X with no abelian subgroup of cardinality $|X|$ and such that X' is finite.

Among the subgroups of X' choose F minimal such that X/F has an abelian subgroup of cardinality $|X|$; let Y/F be such a subgroup so that we have $Y' = F$. Let $Z = Z(Y)$; then $Z \geq Y \cap Z(S)$ and so Y/Z is elementary abelian. Also $|Z| < |Y|$ so that $|Z/Y'| < |Y/Y'|$. Therefore Y has a subgroup W such that $|Y/W| < |Y|$ and $Z \cap W = Y'$. By the minimality of F ($= Y'$) we have $W' = Y'$ and also $|W| = |Y|$, W/W' is elementary abelian, W' is finite elementary abelian and W has no abelian subgroup of cardinality $|W|$.

Unfortunately, this is as far as the reduction will go and we can not always obtain an extraspecial subgroup. The more general form of Theorem 3.12 given by Ehrenfeucht and Faber [20] will give a group G such that G' is finite and elementary abelian of any finite order p^n, G/G' is elementary abelian of cardinality exp $\mathfrak{m}$ and any subgroup H of G such that $|H| = |G|$ satisfies $H' = G'$.

Minimal non-FC-groups

By a minimal non-FC-group we mean a group G which is not an FC-group but in which each proper subgroup is an FC-group.

Attempts to give a complete classification of these groups are certain to fail because of the existence of the Tarski groups (infinite nonabelian groups in which each proper subgroup is finite). However if one imposes conditions which rule out these groups then it is possible to make considerable progress. The results given here have been obtained by Beljaev [7], Beljaev and Sesekin [8] and Bruno and Phillips [13] although some of the proofs given here are rather different.

It is possible to classify those minimal non-FC-groups which have a nontrivial finite factor group. Further results can then be obtained by reducing to this case.

Lemma 8.9. Let H be a normal subgroup of finite index in the minimal non-FC-group G. Then G/H is cyclic of prime power order.

Proof. Since G is not an FC-group there is an element $x \in G$ such that $|G:C_G(x)|$ is infinite. If $\langle H,x\rangle = K$ is a proper subgroup of G then it is an FC-group and so $|K:C_K(x)|$ and hence $|G:C_G(x)|$ would be finite contrary to our choice of x. Thus $G = \langle H,x\rangle$ and G/H is cyclic. If $|G/H| = mn$, where $(m,n) = 1$, then $K_m = \langle H,x^m\rangle$ and $K_n = \langle H,x^n\rangle$ are proper subgroups of G and so are FC-groups. Therefore $|H:C_H(x^m)|$ and $|H:C_H(x^n)|$ are finite and hence $|G:C_G(x^m) \cap C_G(x^n)|$ is finite. But since $(m,n) = 1$, $C_H(x^m) \cap C_H(x^n) = C_H(x)$ and we again have a contradiction to the choice of x. □

Corollary 8.10. Let G^* be the finite residual of the minimal non-FC-group G. If $G^* < G$, then G/G^* is cyclic of prime power order and G^* is a divisible abelian group. □

Theorem 8.11. (Beljaev and Sesekin [8]) Let G^* be the finite residual of the minimal non-FC-group G and suppose that $G^* < G$. Then

(i) $G = \langle G^*, x \rangle$, $x^{p^n} \in G^*$ and $x^p \in Z(G)$,

(ii) G^* is a divisible abelian q-group of finite rank,

(iii) G^* contains no proper infinite subgroup normal in G and $G^* = G'$.

Proof. We have already shown that $G = \langle G^*, x \rangle$, $x^{p^n} \in G^*$ and G^* is a divisible abelian group. The group $\langle G^*, x^p \rangle$ is an FC-group and so, by Theorem 1.9, $G^* \leq Z(\langle G^*, x^p \rangle)$. Thus x^p centralizes G^* and x and hence $x \in Z(G)$.

Now let T be the periodic subgroup of G^* and suppose, if possible, that T is central in G. Let a be any element of infinite order in G^*; then there is an element $b \in G^*$ such that $b^p = a$. Now $\langle b, x \rangle \neq G$ and so $\langle b, x \rangle$ is an FC-group. Therefore $[b,x]$ has finite order and so $[b,x] \in T$. Since T is central in G, we have $[a,x] = [b^p,x] = [b,x]^p = [b,x^p] = 1$. Thus $G^* \leq Z(G)$ contrary to G not being an FC-group. Therefore T is not central in G and so there is a subgroup Q isomorphic to C_{q^∞} such that $Q \not\leq Z(G)$. If $\langle Q,x \rangle < G$, then $\langle Q, x \rangle$ would be an FC-group and, by Theorem 1.9, Q would be centralized by x. Therefore $\langle Q, x \rangle = G$ and G^* is generated by finitely many conjugates of Q. Thus G^* is a q-group of finite rank.

Suppose that $G' < G^*$; then $H = \langle G', x \rangle < G$ and so H is an abelian-by-finite FC-group. It follows that H is centre-by-finite and hence H' is finite. Let $a \in G^*$; then there is an element $b \in G^*$ such that $b^p = a$. Since G/H' is nilpotent of class two, we have

$$[a,x] \equiv [b^p,x] \equiv [b,x]^p \equiv [b,x^p] \equiv 1 \pmod{H'}.$$

Thus G/H' is abelian contrary to G not being an FC-group. Therefore $G^* = G'$ and, since G^* is abelian, each element of G^* is a commutator of the form $[a,x]$. If G^* contains a proper infinite subgroup N normal in G, then $\langle N, x \rangle$ is an FC-group. There is a quasicyclic subgroup Q contained in N and Q must be central. But if $[a,x] \in Q$, then $[a,x]^p = [a,x^p] = 1$ so that

Q must have exponent p. This contradiction completes the proof of part (iii). □.

It will be noted that in a group G satisfying conditions (i) - (iii) a proper subgroup H is abelian if $HG' < G$ and is finite if $HG' = G$, so that Theorem 8.11 does characterize the minimal non-FC-groups with a finite factor group.

We now consider minimal non-FC-groups which are not perfect. Since the groups with a non-trivial finite factor group have been dealt with we can restrict our attention to groups in which G/G' is a divisible abelian group.

Lemma 8.12. Let G be a minimal non-FC-group with no non-trivial finite factor groups. Let G/K be a periodic factor group of G with a homomorphic image G/H isomorphic to C_{p^∞}. Then G/K is a p-group.

Proof. We may write $G = \bigcup_{n=1}^{\infty} H_n$, where H_n/H is cyclic of order p^n. Let P/K be a Sylow p-subgroup of G/K; then $PH = G$. For, otherwise $PH = H_m$ and so P/K is a Sylow p-subgroup of H_n/K for each $n \geq m$, But H_n/K is a periodic FC-group and so, by Theorem 5.4, $PH = H_n > H_m$.

Now let $x \in P$. If $P < G$, then P is an FC-group and so $|P:C_P(x)|$ is finite. Since $P/(P \cap H) \cong C_{p^\infty}$, we have $C_P(x)(P \cap H) = P$ and so $C_G(x)H = PH = G$. Let $H_1 = \langle H,x \rangle$; then $H_1 < G$ and so H_1 is an FC-group. Now $G = C_G(x)H = C_G(x)H_1$ and so $|G:C_G(x)| = |H_1:C_{H_1}(x)|$ which is finite. Since G has no proper subgroups of finite index, we have $x \in Z(G)$. Therefore $P \leq Z(G)$ and so $G = HZ(G)$. It follows that each element of H has only finitely many conjugates in G and, as above, $H \leq Z(G)$. Since $G/H \cong C_{p^\infty}$ this shows that G is abelian and this contradiction shows that $P = G$, as required. □

Theorem 8.13. (Beljaev and Sesekin [8]) Let G be a minimal non-FC-group with $G' < G$. Then G has a non-trivial finite factor group and so satisfies the conditions of Theorem 8.11.

Proof. Suppose that G/G' is a divisible abelian group. It follows from Lemma 8.12 that G/G' is a p-group. As in the proof of Lemma 8.12, we may write $G = \bigcup_{n=1}^{\infty} H_n$. Thus each H_n is an FC-group and so $G' = \bigcup_{n=1}^{\infty} H_n'$ is periodic. Therefore G is a p-group and since G' is an FC-group we have $G'' < G'$. If G/G'' were an FC-group then, since it has no finite factor groups, it would be abelian contrary to $G'' < G'$. Therefore G/G'' is a minimal non-FC-group and we may assume that $G'' = 1$. Let F/G' be a finite subgroup of G/G'; then F is an abelian-by-finite FC-group and so F' is finite. Since G has no finite factor groups, F' is central in G and hence $G' \leq Z(G)$; i.e., G is nilpotent of class two. Let $G/G' = Dr(A_i/G')$, where each A_i/G' is isomorphic to C_{p^∞}. Since G' is central, each A_i is abelian. If $a \in A_j$, then $\langle A_i, a \rangle$ is an FC-group and so a is centralized by A_i. Therefore $[A_i, A_j] = 1$ and G is abelian contrary to $G'' < G'$. □

The classification of minimal non-FC-groups therefore reduces to considering perfect groups with no non-trivial finite factor groups. In order to avoid the Tarski groups, Bruno and Phillips suggest that one should consider locally graded groups. A group G is *locally graded* if each non-trivial finitely generated subgroup of G has a non-trivial finite homomorphic image.

Lemma 8.14. (Bruno and Phillips) A locally graded minimal non-FC-group is locally finite. Any minimal non-FC-group is countable.

Proof. Let x be an element with $|G:C_G(x)|$ infinite. Then there is a countable subgroup H such that $|H:C_H(x)|$ is infinite. If G is a minimal non-FC-group then H must be equal to G.

Now let G be locally graded. Then G is not finitely generated and so has a local system of finitely generated proper subgroups $X_i, i \in I$. Each X_i is an FC-group and so X_i' is finite. If G is perfect then $G = G'$ is locally finite. If G is not perfect then we can use Theorem 8.13. □

Theorem 8.15. (Bruno and Phillips) A locally graded minimal non-FC-group which is perfect and locally soluble is a p-group.

Proof. Let U/V be a p-chief factor of G. Since G/V is a minimal non-FC-group we may assume that V = 1 so that U is a minimal normal subgroup of G. Since G is countable and locally finite it has a Sylow p-subgroup P and a Sylow p′-subgroup Q such that PQ = G. Consider UQ. If G = UQ, then G/U is an FC-group and we have either a finite or an abelian factor group. Thus

$UQ < G$ and so UQ is an FC-group. If $P < G$ and $x \in U$, then x^P is finite and so $x^G = (x^P)^{UQ}$ is finite. Thus $x \in Z(G)$ and U is central. If $P = G$, then again U is central. Therefore G is locally nilpotent and so a product of its Sylow subgroups. The minimality shows that G is a p-group. □

Rather than minimal non-FC-groups, Beljaev and Sesekin actually considered minimal non-$\mathfrak{F}\mathfrak{A}$-groups and Beljaev [7] showed that there are no locally finite perfect minimal non-$\mathfrak{F}\mathfrak{A}$-groups so that the locally graded minimal non-$\mathfrak{F}\mathfrak{A}$-groups are completely characterized by Theorem 8.11. Beljaev's work involves the consideration of certain infinite simple groups which might occur and makes use in particular of the locally finite version of Bender's theorem (see [60] , p.142). Those of you who have read this far will probably agree that that type of argument would be rather out of place in these notes.

We end by noting the most obvious gap in this classification.

Question 8A. Is every locally graded minimal non-FC-group given by the characterization in Theorem 8.11?

Exercises

1. An FC-group with minimal condition on normal subgroups or abelian subgroups is a Černikov group. [For minimal condition on abelian subgroups use Theorem 7.20.]

2. An FC-group with minimal condition on centralizers (in particular, a linear FC-group) is centre-by-finite.

3. An FC-group with maximal condition on normal subgroups or abelian subgroups is centre-by-finite and so satisfies the maximal condition on all subgroups.

4. (Fedorov [30]) If G is an infinite group in which each non-trivial subgroup has finite index then G is cyclic.

5. (Černikov [16]) G is an FC-group if and only if G has a torsion-free central subgroup K such that G/K is a periodic FC-group. [Show that a finitely generated subgroup F of G has finite derived subgroup and hence $C_G(F) = C_G(FK/K)$.]
   If G = HZ(G) and H is an FC-group then G is an FC-group. However, in any class two nilpotent group, G/Z(G) is an FC-group.

6. If N is an infinite normal subgroup of the (locally nilpotent)-by-finite FC-group G, then $N \cap Z(G) \neq 1$.

7. Let G be an FC-group with $|G'| < \mathfrak{m}$. Then $|G/Z_2(G)| < \mathfrak{m}$. [Choose $N \triangleleft G$ such that $N \cap G'Z = Z$ and consider $Y = C_G(G/N)$.]

8. The join of any set of ascendant subgroups of an FC-group G is an ascendant subgroup. However the join of two subnormal subgroups need not be subnormal. For a counterexample here one needs finite groups G_n containing subnormal subgroups H_n and K_n each having defect 3 such

that $J_n = \langle H_n, K_n \rangle$ has defect n in G_n; then let $G = \mathrm{Dr}_{n=1}^{\infty} G_n$.

9. If the FC-group G is the product of two locally nilpotent subgroups then G is locally soluble. [Consider the finite factor groups.]

10. Let G be an FC-group. The set of left-Engel elements

$$\{x \in G: \text{ for each } g \in G \text{ there is an integer } n \text{ such that } [g,_n x] = 1\}$$

coincides with the locally nilpotent radical of G. [Consider the finitely generated normal subgroups of G.]
The set of right-Engel elements

$$\{x \in G: \text{ for each } g \in G \text{ there is an integer } n \text{ such that } [x,_n g] = 1\}$$

coincides with the hypercentre of G. [Consider the finite factor groups of G.]

11. (Trahtenberg [109]) Let $\Phi(G)$ be the Frattini subgroup of the FC-group G and let $N \lhd G$. Then N is locally nilpotent (locally supersoluble) if and only if $N/(N \cap \Phi(G))$ is locally nilpotent (locally supersoluble). [Consider the finite factor groups.]

12. (Neumann [74]) The Sylow p-subgroups of any FC-group G (not necessarily periodic) are locally conjugate in G.

13. An FC-group G is complemented (that is, every subgroup H has a complement K such that $HK = G$, $H \cap K = 1$) if and only if H is a subgroup of a direct product of finite groups of squarefree order.
[$G = Z(G) \times M$ where $Z(M) = 1$ and so $M \in \mathcal{SDF}$, then use the result for finite groups.]

14. (Nishigôri [79]) The isolator of a subgroup H is the set

$$I(H) = \{x \in G: x^n \in H \text{ for some integer } n\}.$$

If H is a subgroup of an FC-group G, then I(H) is a subgroup. [$I(H) = I(H \cap Z) = I((H \cap Z)G')$ and $I((H \cap Z)G')/(H \cap Z)G'$ is the periodic subgroup of $G/(H \cap Z)G'$.]

15. (Tkačenko [96a]) If the periodic FC-group G has a Sylow p-subgroup P such that $P \lessapprox Z(N_G(P))$, then G has a normal Sylow p′-subgroup. [Consider the finite factor groups.]

16. (Tkačenko [96a]) If $N_G(H)/C_G(H)$ is a p-group for each p-subgroup H of the periodic FC-group G, then G has a normal Sylow p′-subgroup. [Consider the finite normal subgroups.]

17. Let $K = \{S_{p'}\}$ be a Sylow complement system of the locally soluble periodic FC-group G. The prefrattini subgroup of G associated with K is defined to be

$$W = \bigcap \{M : M \text{ is a maximal subgroup of } G \text{ and } M \geq S_{p'}, \text{ for some } p\}.$$

The prefrattini subgroups of G avoid the complemented chief factors of G and cover the rest. [Consider the finite factor groups.]

18. (Haimo [48]) Let G be a group with $|G'| = n$, finite. Then $|G/Z_2(G)| = m$ is finite and the mapping $x \to x^{2mn}$ is a central endomorphism of G. [Using the transfer in G/Z, $(xy)^m = x^m y^m z(x,y)$, where $z(x,y) \in Z \cap G'$.
Therefore $(xy)^{2mn} = x^{2mn} y^{2mn} [y^m, x^m]^{\binom{2n}{2}} z(x,y)^{2n} = x^{2mn} y^{2mn}$.]

19. If a group G is the union of finitely many nilpotent subgroups $H_1, \ldots, H_k$ each of class c, then $Z_{k(c-1)+1}(H)$ has finite index in G.

References

Items marked with an asterisk are in Russian.

1. I.N. Abramovskiĭ, *On subgroups of direct products of finite groups. IX th All-Union Algebraic Colloquium, Gomel' (1968), 3-4.

2. I.N. Abramovskiĭ, *Subgroups of direct products of groups. "Problems in the theory of groups and rings." In-t Fiziki SO AN SSSR, Krasnoyarsk (1973), 3-8.

3. R. Baer, Sylow theorems for infinite groups. Duke Math. J. 6 (1940), 598-614.

4. R. Baer, Finiteness properties of groups. Duke Math. J. 15 (1948), 1021-1032.

5. R. Baer, Endlichkeitskriterien für Kommutatorgruppen. Math. Ann. 124 (1952), 161-177.

6. R. Baer, Automorphismengruppen von Gruppen mit endlichen Bahnen gleichmässig beschränkte Mächtigkeit. J. Reine Angew. Math. 262/3 (1973), 93-119.

7. V. V. Beljaev, *Groups of Miller-Moreno type. Sibirskiĭ Mat. Ž. (1978), 509-514.

8. V. V. Beljaev and N. F. Sesekin, *On infinite groups of Miller-Moreno type. Acta Math. Acad. Sci. Hungar. 26 (1975), 369-376.

9. M. L. Berlinkov, *On the lattice of subgroups of layer-finite groups. Uspehi Mat. Nauk 12 (1957), 267-271.

10. N. Bourbaki, Theory of Sets. Addison-Wesley 1968.

11. N. Bourbaki, General topology, Part 1. Addison-Wesley 1966.

12. I. M. Bride, Second nilpotent BFC-groups. J. Austral. Math. Soc. 11 (1970), 9-18.

13. B. Bruno and R.E. Phillips, Minimal non- FC-groups. Abstracts Amer. Math. Soc. 2 (1980), 565.

14. S. N. Černikov, *On the complementation of the Sylow π-subgroup in certain classes of infinite groups. Mat. Sb. 37 (1955), 557-566.

15. S. N. Černikov, *On groups with finite classes of conjugate elements Dokl. Akad. Nauk SSSR 114 (1957), 1177-1179.

16. S. N. Černikov, *On the structure of groups with finite classes of conjugate elements. Dokl. Akad. Nauk SSSR 115 (1957), 60-63.

17. S. N. Černikov, *On layer-finite groups. Mat. Sb. 45 (1958), 415-416.

18. E. C. Dade, Carter subgroups and Fitting heights of finite soluble groups. Illinois J. Math. 13 (1969), 449-514.

19. A. P. Dicman, *On p-groups. Dokl. Akad. Nauk SSSR 15 (1937), 71-76.

20. A. Ehrenfeucht and V. Faber, Do infinite nilpotent groups always have equipotent abelian subgroups? Kon. Nederl. Akad. Wet. A75 (1972 202-209.

21. J. Erdös, The theory of groups with finite classes of conjugate elements. Acta. Math. Acad. Sci. Hungar. 5 (1954), 45-58.

22. P. Erdös, A. Hajnal and R. Rado, Partition relations for cardinal numbers. Acta. Math. Acad. Sci. Hungar. 16 (1965), 93-196.

23. I. I. Eremin. *Groups with finite classes of conjugate abelian subgroups. Mat. Sb. 47 (1959), 45-54.

24. I. I. Eremin, *On central extensions by means of thin layer-finite groups. Izv. Vysš. Uč. Zaved. Matematika 2 (1960), 93-95.

25. I. I. Eremin, *Groups with finite classes of conjugate infinite subgroups. Perm. Gos. Univ. Učen. Zap. 17 (1960), 13-14.

26. I. I. Eremin, *On groups with finite classes of conjugate subgroups with a given property. Dokl. Akad. Nauk SSSR 137 (1961), 772-773.

27. V. Faber, Large abelian subgroups of some infinite groups. Rocky Mountain J. Math. 1 (1971), 677- 685.

28. V. Faber, R. Laver and R. McKenzie, Coverings of groups by abelian subgroups. Canad. J. Math. 30 (1978), 933-944.

29. V.Faber and M. J. Tomkinson, On theorems of B. H. Neumann concerning FC-groups, II. Rocky Mountain J. Math.13 (1983), 495-506.

30. Ju. G. Fedorov, *On infinite groups every non-trivial subgroup of which has finite index. Uspehi Mat. Nauk 6 (1951), 187-189.

31. U. Felgner, On $\aleph_0$-categorical extraspecial p-groups. Logique et Analyse 71/72 (1975), 408-428.

32. U. Felgner, The model theory of FC-groups. "Mathematical Logic in Latin America" Studies in Logic, Vol.99. North Holland (1980), 163-190.

33. B. Fischer, W. Gaschütz and B. Hartley, Injektoren der endlichen auflösbaren Gruppen. Math. Z. 102 (1967), 337-339.

34. L. Fuchs, On groups with finite classes of isomorphic subgroups. Publ. Math. Debrecen 3 (1954), 243-252.

35. L. Fuchs, Infinite abelian groups, Vol.I. Academic Press 1970.

36. M. Garcia Esnaola, Inyectores en grupos localmente finitos-resolubles. Tesis Doctoral (Universidad de Zaragoza, 1979).

37. A. D. Gardiner, B. Hartley and M. J. Tomkinson, Saturated formations and Sylow structure in locally finite groups. J. Algebra 17 (1971), 177-211.

38. W. Gaschütz, Zur Theorie der endlichen auflösbaren Gruppen. Math. Z. 80 (1963), 300-305.

39. A. V. Gel'fand, *Some estimates for the index of the centre in an FIZ-group. "Group theory and some questions in algebra." Permskiĭ Gos. In-ta 1975, 29-45.

40. P. A. Gol'berg, *Sylow π-subgroups of locally normal groups. Mat. Sb. 19 (1946), 451-460.

41. Ju. M. Gorčakov, *On embedding of locally normal groups in direct products of finite groups. Dokl. Akad. Nauk SSSR 138 (1961), 26-28.

42. Ju. M. Gorčakov, *On locally normal groups. Mat. Sb. 67 (1965), 244-254.

43. Ju. M. Gorčakov, *Locally normal groups. II All-Union Symposium on Group Theory (Batum 1966). Tbilisi 1967, 10-12.

44. Ju. M. Gorčakov, *Locally normal groups. Sibirskiĭ Mat. Ž. 12 (1971), 1259-1272.

45. Ju. M. Gorčakov, *Theorems of Prüfer-Kulikov type. Alg. i Log. 13 (1974), 655-661.

46. Ju. M. Gorčakov, *Subgroups of direct products. Alg. i Log. 15 (1976), 622-627.

47. Ju. M. Gorčakov, *Groups with finite classes of conjugate elements. Sovremennaja Algebra, Nauka 1978.

48. F. Haimo, Groups with a certain condition on conjugates. Canad. J. Math. 4 (1952), 309-321.

49. P. Hall, Periodic FC-groups. J. London Math. Soc.34 (1959), 289-304.

50. B. Hartley, Some examples of locally finite groups. Arch. der Math. 23 (1972), 225-231.

51. B. Hartley, Subgroups of locally normal groups. Compositio Math. 32 (1976), 185-201.

52. B. Hartley, Profinite and residually finite groups. Rocky Mountain J. Math. 7 (1977), 193-217.

53. B. Hartley and J. R. Parker, Local conjugacy classes and Baire's theorem. Arch. der Math. 27 (1976), 341-346.

54. P. Higgins, An introduction to topological groups. LMS Lecture Note Series, No.15. Cambridge U.P. 1974.

55. B. Huppert, Endliche Gruppen, I. Springer 1967.

56. I. Kaplansky, An introduction to differential algebra. Herman 1951.

57. I. Kaplansky, Linear algebra and geometry: a second course. Allyn and Bacon 1969.

58. M. I. Kargapolov, *On conjugacy of Sylow p-subgroups of a locally normal group. Uspehi Mat. Nauk 12 (1957), 297-300.

59. M. I. Kargapolov, *On the theory of semi-simple locally normal groups. Naucn Dokl. Vysš. Fiz. Mat. Nauk 6 (1958), 3-7.

60. O. H. Kegel and B.A. F. Wehrfritz, Locally finite groups. North Holland 1973.

61. L. A. Kurdačenko, *FC-groups with layer-finite periodic part. IV th All-Union Symposium on Group Theory, Novosibirsk 1973, 93-98.

62. L. A. Kurdačenko, *Nonperiodic groups with a bound on the layers of elements. Ukr. Mat. Ž. 26 (1974), 386-389.

63. L. A. Kurdačenko, *FC-groups with layer-finite periodic part. "Some problems in group theory." In-t Matem. AN USSR, Kiev 1975, 160-172.

64. L.A. Kurdačenko, *FC-groups with a bound on the orders of elements of the periodic part. Sibirskiĭ Mat. Ž. 16 (1975), 1205-1213.

65. L. A. Kurdačenko, *FC-groups whose periodic part can be embedded in a direct product of finite groups. Mat. Zametki 21 (1977), 9-20.

66. A. G. Kuroš, The theory of groups. Chelsea 1960.

67. I. D. Macdonald, A class of FC-groups. J. London Math. Soc. 34 (1959) 73-80.

68. I. D. Macdonald, Some explicit bounds in groups with finite derived groups. P. London Math. Soc. 11 (1961), 23-56.

69. H. Meyn, FC-groups and related classes. Rend. Sem. Mat. Padova 47 (1972), 65-75.

70. B. H. Neumann, Groups with finite classes of conjugate elements. P. London Math. Soc. 1 (1951), 178-187.

71. B. H. Neumann, Groups covered by permutable subsets. J. London Math. Soc. 29 (1954), 236-248.

72. B. H. Neumann, Groups covered by finitely many cosets. Publ. Math. Debrecen 3 (1954), 227-242.

73. B. H. Neumann, Groups with finite classes of conjugate subgroups. Math. Z. 63 (1955), 76-96.

74. B. H. Neumann, Isomorphism of Sylow subgroups of infinite groups. Math. Scand. 6 (1958), 299-307.

75. B. H. Neumann, Questions and examples on group coverings. Acta Math. Acad. Sci. Hungar. 26 (1975), 291-293.

76. B. H. Neumann, A problem of Erdös on groups. J. Austral. Math. Soc. 21 (1976), 467-472.

77. P. M. Neumann, An improved bound for BFC p-groups. J. Austral. Math. Soc. 11 (1970), 19-27.

78. P. M. Neumann and M. R.Vaughan-Lee, An essay on BFC-groups. P. London Math. Soc. 35 (1977), 213-237.

79. N. Nishigôri, On some properties of FC-groups. J. Sci. Hiroshima Univ. Ser. A-I Math. 21 (1957/8), 99-105.

80. J. R. Parker, A topological approach to a class of residually finite groups. Ph.D. Thesis (University of Warwick, 1973).

81. Ja. D. Polovickiĭ, *Layer-extremal groups. Mat. Sb. 56 (1962), 95-106.

82. Ja. D. Polovickiĭ, *Locally extremal and layer-extremal groups. Mat. Sb. 58 (1962), 685-694.

83. D. J. S. Robinson, On the theory of groups with extremal layers. J. Algebra 14 (1970), 182-193.

84. D. J. S. Robinson, Finiteness conditions and generalized soluble groups. Parts 1 and 2. Ergebnisse der Math. Bde. 62/63. Springer 1972.

85. D. J. S. Robinson, S. E. Stonehewer and J. Wiegold, Automorphism groups of FC-groups. Arch. der Math. 40 (1983), 401-404.

86. I. Schur, Über die Darstellung der endlichen Gruppen durch gebrochene lineare Substitutionen. J. Reine Angew Math. 127 (1904), 20-50.

87. L. Schiefelbusch, The Trofimov number of some infinite groups with finiteness conditions. Arch. der Math. 18 (1967), 122-127.

88. W. R. Scott, On a result of B.H. Neumann. Math. Z. 66 (1956), 240.

89. T. Ja. Semenova, *On nilpotent subgroups of FC-groups. "Some problems in the theory of groups and rings." In-t Fiz. SO AN SSSR, Krasnoyarsk (1973), 150-159.

90. T. Ja. Semenova, *On the relation between classes of locally conjugate subgroups and conjugate subgroups. All-Union Algebraic Symposium, Gomel' (1975), 162.

91. S. Shelah, Classification theory and the number of non-isomorphic models. Studies in Logic, Vol.92. North Holland 1978.

92. J. A. H. Shepperd and J. Wiegold, Transitive permutation groups and groups with finite derived groups. Math. Z. 81 (1963), 279-285.

93. S. E. Stonehewer, Locally soluble FC-groups. Arch. der Math. 16 (1965), 158-177.

94. S. E. Stonehewer, Some finiteness conditions in locally soluble groups. J. London Math. Soc. 43 (1968), 689-694.

95. S. E. Stonehewer, Automorphisms of locally nilpotent FC-groups. Math. Z. 148 (1976), 85-88.

96. W. W. Stothers and M. J. Tomkinson, On infinite families of sets. Bull. London Math. Soc. 11 (1979), 23-26.

96a. A. N. Tkačenko,*Burnside and Frobenius theorems in a class of locally normal groups. "Constructive description of groups with properties prescribed on their subgroups. Kiev (1980), 71-76.

96b. A. N. Tkačenko,*On locally normal groups with complementable normal subgroups. Ukrain. Mat. Ž. 33 (1981), 557-558.

97. S. Tôgô, Ascendancy in locally finite groups. Hiroshima Math. J. 12 (1982), 93-102.

98. M. J. Tomkinson, Generalized formations of locally soluble FC-groups. Ph.D. Thesis (University of Newcastle upon Tyne, 1968).

99. M. J. Tomkinson, Formations of locally soluble FC-groups. P. London Math. Soc. 19 (1969), 675-708.

100. M. J. Tomkinson, Local conjugacy classes. M. Zeit. 108 (1969), 202-212.

101. M. J. Tomkinson, Local conjugacy classes, II. Arch. der Math. 20 (1969), 567-571.

102. M. J. Tomkinson, $\mathfrak{F}$-injectors of locally soluble FC-groups. Glasgow Math. J. 10 (1969), 130-136.

103. M. J. Tomkinson, Extraspecial sections of periodic FC-groups. Compositio Math. 31 (1975), 285-302.

104. M. J. Tomkinson, Residually finite periodic FC-groups. J. London Math. Soc. 16 (1977), 221-228.

105. M. J. Tomkinson, On the commutator subgroup of a periodic FC-group. Arch. der Math. 31 (1978), 123-125.

106. M. J. Tomkinson, A characterization of residually finite periodic FC-groups. Bull. London Math. Soc. 13 (1981), 133-137.

107. M. J. Tomkinson, On theorems of B. H. Neumann concerning FC-groups. Rocky Mountain J. Math. 11 (1981), 47-58.

108. M. J. Tomkinson, A generalization of FC-groups. Arch. der Math.

109. A. M. Trahtenberg, *The Frattini subgroup of an FC-group. Matem. Issled 7 (1972), 248-252.

110. J. Wiegold, Groups with boundedly finite classes of conjugate elements. Proc. Roy. Soc. London Ser. A 238 (1957), 389-401.

111. N. H. Williams, Combinatorial set theory. Studies in Logic, Vol.91. North Holland 1977.

Additional References

112. B. Bruno and R.E. Phillips. On minimal conditions related to Miller-Moreno type groups. Rend. Sem. Mat. Padova 69 (1983), 153-168.

113. L. A. Kurdačenko. *Embeddability of an FC-group into a direct product of finite groups and an abelian torsion-free group. Mat. Zametki 29 (1981), 359-373.

114. L. A. Kurdačenko. *FC-groups with bounded periodic part. Ukrain. Mat. Ž. 35 (1983), 374-378.

Index